Studies in Computational Intelligence

Volume 1095

Series Editor

Janusz Kacprzyk, Polish Academy of Sciences, Warsaw, Poland

The series "Studies in Computational Intelligence" (SCI) publishes new developments and advances in the various areas of computational intelligence—quickly and with a high quality. The intent is to cover the theory, applications, and design methods of computational intelligence, as embedded in the fields of engineering, computer science, physics and life sciences, as well as the methodologies behind them. The series contains monographs, lecture notes and edited volumes in computational intelligence spanning the areas of neural networks, connectionist systems, genetic algorithms, evolutionary computation, artificial intelligence, cellular automata, self-organizing systems, soft computing, fuzzy systems, and hybrid intelligent systems. Of particular value to both the contributors and the readership are the short publication timeframe and the world-wide distribution, which enable both wide and rapid dissemination of research output.

Indexed by SCOPUS, DBLP, WTI Frankfurt eG, zbMATH, SCImago.

All books published in the series are submitted for consideration in Web of Science.

Massimo Buscema · Masoud Asadi-Zeydabadi ·
Giulia Massini · Weldon A. Lodwick ·
Marco Breda · Riccardo Petritoli ·
Francis Newman · Francesca Della Torre

The Topological Weighted Centroid: A New Vision of Geographic Profiling

Theory and Applications

Springer

Massimo Buscema
Semeion Centro Ricerche di Scienze della Comunicazione
Rome, Italy

Giulia Massini
Semeion Centro Ricerche di Scienze della Comunicazione
Rome, Italy

Marco Breda
Semeion Centro Ricerche di Scienze della Comunicazione
Rome, Italy

Francis Newman
Department of Radiation Oncology
University of Colorado Denver
Denver, CO, USA

Masoud Asadi-Zeydabadi
Department of Physics
University of Colorado Denver
Denver, CO, USA

Weldon A. Lodwick
Department of Mathematical and Statistical Sciences
University of Colorado Denver
Denver, CO, USA

Riccardo Petritoli
Semeion Centro Ricerche di Scienze della Comunicazione
Rome, Italy

Francesca Della Torre
Semeion Centro Ricerche di Scienze della Comunicazione
Rome, Italy

ISSN 1860-949X ISSN 1860-9503 (electronic)
Studies in Computational Intelligence
ISBN 978-3-031-28900-2 ISBN 978-3-031-28901-9 (eBook)
https://doi.org/10.1007/978-3-031-28901-9

This Springer imprint is published by the registered company Springer Nature Switzerland AG
The registered company address is: Gewerbestrasse 11, 6330 Cham, Switzerland

Preface

This monograph describes in detail the foundations and applications of a new approach to geographic profiling called Topological Weighted Centroid. Both example applications and small simple illustrative examples are provided. This monograph is designed for those interested in the dynamics of geographical processes such as criminality/terrorist activity, the evolution of a disease in time and space, archeological dynamics, environmental issues, and political/cultural processes. The basic underlying principles center around looking at geographical relationships from a free energy and entropy point of view. One of the algorithms couples the analysis with graphs and minimal spanning trees and another couples the geographical analysis with an artificial adaptive system (Auto-Contractive Maps).

Associated with the algorithms presented are *C* and *Matlab* codes. The *Matlab* code is found in the Appendix. The *C* code can be downloaded from *https://github.com/SemeionResearch/TWC*.

Rome, Italy	Massimo Buscema
Denver, USA	Masoud Asadi-Zeydabadi
Rome, Italy	Giulia Massini
Denver, USA	Weldon A. Lodwick
Rome, Italy	Marco Breda
Rome, Italy	Riccardo Petritoli
Denver, USA	Francis Newman
Rome, Italy	Francesca Della Torre

Contents

Chapter 1
Introduction

Topological Weighted Centroid (TWC) is a set of algorithms that take a distribution of locations whose 2-dimensional (latitude/longitude) geographical coordinates are known input data and extracts patterns, dynamics, that is, information. In particular, TWC imputes the past "source" that caused the current distribution of data and its future configurations. By "source" we mean *the location of the place, with respect to minimal energy, from which the given distribution of the input data of the geographical event in question would be attained.* TWC is an intelligent geographical information system method that has been outlined in the literature and can be found in, for example, [1]. This data driven approach was developed based on statistical thermodynamics, diffusion theory, optimizing free energy and entropy. Generally speaking, the TWC algorithms identify the center locations from which the impetus of current configuration of data arose or will arise. These locations are calculated as past sites, present sites, near future sites, and longer future sites. This monograph uses the word "phenomenon" to denote various events that occur across geographical space (disease, terror attacks, robberies, drug distribution, and so on).

There are four sets of TWC algorithms [1]: (1) *TWC-Original*, (2) *TWC-Frequency*, (3) *TWC-Sematic*, and (4) *TWC-Windowing*. Each of the four types of TWC have five algorithms that can be applied. The five TWC algorithms are called *TWC-Alpha*, *TWC-Beta*, *TWC-Gamma*, *TWC-Theta,* and *TWC-Iota*.

This monograph deals with maps defined by a discrete set of points in a two dimensional space of real numbers $\mathbb{R}^2$, a rectangular grid. The number of points in the grid is determined by the desired or required resolution of the analysis. There is also a given distribution of points, the input data, in a subset of the grid points, most often a proper subset of the set of all grid points of the rectangular area of interest. Square areas are chosen to simplify the subscript notation. Thus we define two sets of points as follows.

M. Buscema et al., *The Topological Weighted Centroid: A New Vision of Geographic Profiling*, Studies in Computational Intelligence 1095,
https://doi.org/10.1007/978-3-031-28901-9_1

Definition 1 Given a two-dimensional rectangle geographic region, we call a **map** $\Re$ at a given $M \times N$ resolution, the set of MN points on this grid is called **grid points** and denoted

$$\Re = \left(x_i, y_j\right), 1 \leq i \leq M, 1 \leq j \leq N. \quad (1.1)$$

Definition 2 Given a map $\Re$, a subset $U \subset \Re$ of $K \leq MN$ points, called a **set of distribution points** or the **assigned points** or the **data points** (the input data set), is the set

$$U = (x_k, y_l) \in \Re, k \subset \{1, ..., M\}, l \subset \{1, ...N\},$$

and most often $K << MN$.

Reference

1. M. Buscema, G. Massini, P.L. Sacco, The topological weighted centroid (TWC): a topological approach to the time-space structure of epidemic and pseudo-epidemic processes. Physica A: Stat. Mech. Appl. **492**, 582–627 (2018)

Chapter 2
Precursor to TWC

A complementary approach to the topological one of TWC is what is called "Geographic Profiling", which came out of forensic research (see, for example, [1–7]), although geographic profiling was also adapted and applied to different kinds of problems and policy concerns. In addition, it has been applied to detect the outbreak points of many epidemics (see [6]), as well as to analyze other biologically-motivated problems such as the spatial patterns of animal foraging (see [7]). For geographic profiling as applied to epidemics, for example, researchers try to determine the outbreak area of an epidemic and its likely patterns of spatial diffusion, making use of only one snapshot of the past diffusion history of the epidemics. We call this approach of geographic profiling, the *geometric approach* to distinguish it from the *topological approach* of TWC. The advantage of the geometric approach is that it needs relatively little data and, additional data, as it becomes available, they can be quickly and easily added to update estimations.

However, the geometric approach also has substantial limitations as follows.

1. Its validation protocol is statistically weak.
2. The estimated outbreak area depends on the spatial concentration of the event locations. If observed locations are relatively concentrated, the probability of finding the outbreak increases. But when the spatial pattern is more scattered, the method could arrive at ill-founded conclusions.
3. Consequently, the outbreak area estimated by this method generally lies within the convex hull of the distribution of the event locations, and this is a major methodological shortcoming which seriously limits its applicability.

Regardless of these drawbacks, this approach has been applied very recently to many real cases and it is continually updated with new algorithms, very often inspired by the theory of conditional probability (see [8–11]). Thus we have at least two methods to analyze problems occurring in space.

M. Buscema et al., *The Topological Weighted Centroid: A New Vision of Geographic Profiling*, Studies in Computational Intelligence 1095,
https://doi.org/10.1007/978-3-031-28901-9_2

1. *Geometric*—This is a "top down" (model driven) method which will be more fully described in Sect. 2.1 below.
2. *Topological*—This is a "bottom up" (data driven) method which will be more fully described in the Sect. 2.2 below.

There is another approached that we classify as *algebraic*. These methods essentially consist of differential equation models both partial and ordinary. Partial differential equations have the ability to encode dynamics and location simultaneously. We will not discuss algebraic methods, since their approach is significantly different than what is presented here.

Broadly speaking, systems that deal with geographic data are called *Geographic Information Systems*. These systems provide computerization of storage, display, and analytical processing of geographic data. Since TWC stores, processes (analyzes), and displays geographic data, it is a geographic information system. In addition, it adds an intelligent component to the analysis of geographic data (neural network processing) transforming input data as well as projecting and drawing conclusions such interconnections or pathways (graphs in the mathematical sense) over which the phenomenon occurs. Hence, TWC can be considered as an intelligent geographic information system. Algebraic methods such as partial differential equations are distinct from geographic information systems.

2.1 Theory of Location and Geographic Profiling—Geometric Systems

This theory attempts to find an optimal location for any particular distribution of activities related to population or events over a region according to a specific criterion. One could also reverse this logic. Given the distribution of demand, the theory could be applied to estimate a central location from which travel distance or time is minimized. The central location from which travel distance and/or time is minimized is called the *anchor point*.

Many algorithms were invented to estimate the anchor point in geographic profiling. One possible use is the estimation of a serial offender's anchor point, based on knowledge of the locations of the offender's crime sites. We believe that the anchor point, as previously defined, and the traditional center of mass are not the most interesting points in location theory. Every agent (predator or virus for example) also has a defined quantity of energy to expend. This energy is consumed during travel (such as virus expansion or hunting). The energy consumption can determine the length of the hunt or the strength of the virus' expansion. Other random incidents, such as physical or psychological constraints, can also interrupt travel.

Definitely, we have to consider not only distances, but also the energy expenditure, and the geographical and contextual constraints of the territory. We define *operation point*, also called the *hidden point,* the point which minimizes the distances and the energy expenditure in case of success and failure. More generally, to estimate the hidden point of any spatial distribution of events we have to consider at least the following.

1. The specific metric of the space such as the Euclidean distance, block distance, travel distance, travel time, energy usage, and so on.
2. A distance decay function from the known sites or "centers".
3. Geographic factors and idiosyncratic style of action, minimization of energy in case of a random series of failures.

We can rewrite the operation point problem as a probabilistic problem of maximization of a likelihood function (see [3]),

$$P(x, z) = F(d(x, z))G(x)N(x, z)$$

where

$$\begin{aligned} d(\circ) &= \text{a specific metric,} \\ F(\circ) &= \text{a specific decay distance function,} \\ G(\circ) &= \text{a function considering the geographic factors,} \\ N(\circ) &= \text{a function to normalize the operation point in} \\ &\quad \text{case of random failures.} \end{aligned}$$

Most of the algorithms presented in the literature only focus on the metric of the space and the shape of the decay distance function. Following this approach, the operation point is located in a region with a high "Hit Score":

$$z = \arg\max_x \{S(y_k)\}$$

$$S(y_k) = \sum_i^N F(d(y_k, x_i));$$

where

$$\begin{aligned} z &= \text{operation point,} \\ F(\circ) &= \text{distance decay function,} \\ d(\circ) &= \text{specific metric,} \\ x_i &= \text{assigned sites,} \\ y_k &= \text{an operational point candidate,} \\ N &= \text{number of sites,} \\ S(\circ) &= \text{score function.} \end{aligned}$$

There are two main geographic profiling algorithms.

1. The Rossmo algorithm (see [2]). This algorithm has been used in numerous geographic profiling analyses and it can be represented by a simple equation.

$$p_{ij} = p\sum_{n=1}^{M}\left[\frac{\phi_{ij}}{\left(|X_i-x_n|+|Y_j-y_n|\right)^f}+\frac{(1-\phi_{ij})B^{g-f}}{\left(2B-|X_i-x_n|-|Y_j-y_n|\right)^g}\right] \tag{2.1}$$

where

$$\phi_{ij} = \begin{Bmatrix} 1 \text{ if } \left(|X_i - x_n| + |Y_j - y_n|\right) > B \\ 0 \quad \text{otherwise} \end{Bmatrix},$$

M = total number of distribution points,
n = index of each distribution point,
i, j = indices of each grid point,
X, Y = coordinate of each grid point,
x, y = coordinate of each distribution point,
B = buffer zone,
f, g = parameters that need to be fixed according to the data,
p = parameter, determined experimentally for p_{ij} to be a probability.

2. The Canter algorithm (see [5, 12]). This algorithm also is very well known and used in the geographic profiling literature. Its basic equation is very simple.

$$p_{ij} = \sum_{n=1}^{M}\alpha e^{-\beta\sqrt{(X_i-x_n)^2+(Y_i-y_n)^2}} \tag{2.2}$$

where

α, β = free parameters adapted to the data set.

2.2 Topological Systems

A new approach that has appeared relatively recently is the *topological approach*. It shares many characteristics of the geometrical approach, and some of the differential-algebraic approach as well. However, the new approach makes it possible to estimate the location of both the outbreak point and the dynamic unfolding of the phenomenon by working only on the spatial distribution of the events, without any reference to the chronology and frequency of the observations. That is, the topological approach is based purely on the spatial characteristics of the phenomenon. Moreover, unlike the two previous approaches, the topological one works on the basis of an optimization principle, that is, it adaptively explores the fitness landscape with respect of certain quantities (free energy and entropy), and determines the optimal solution accordingly. The aim of this section is to introduce the theoretical foundations of this topological approach in a systematic manner, and to illustrate its potential through its application to a few real case studies. Since the topological approach does not need to take into account the actual sequence and frequency of observations to derive the structural features of the underlying process and to predict its future unfolding, it is possible to apply it not only to other kinds of diffusion phenomena, as it was the case for the geographical profiling approach, but also to the analysis of cases where the spatial distribution of the events was not the outcome of a diffusion process in the proper sense of the word. For instance, the topological method can be applied to characterize the structural features of the distribution of geographical location of points that describe a certain type of human activity (urban settlements, social and economic activities, and so on), which were not actually shaped by a diffusion process of a different phenomenon, or more generally by the action of a well-recognizable agent, but which nevertheless yield an observed spatial pattern that is observationally indistinguishable from one generated by an epidemic or an agent-driven one. We speak in this case of a pseudo-diffusion pattern or, in case we focus upon the processes of the phenomenon as the conceptual benchmark, of a pseudo-phenomenon one.

It will be shown that the topological approach to spatial distributions of human activities such as epidemic patterns allows us to gain considerable insight into their structural characteristics, as well as to effectively forecast their future unfolding. The topological approach, therefore, has a vast range of application to diverse classes of problems where the spatial distribution dimension has a key relevance.

2.3 Comments

Understanding the spatial and temporal structure of geographic events is a theme of major concern in scientific as well as policy-making institutions. Properly tracking events that occur in geographic regions is a formidable challenge that calls for a comprehensive, multidisciplinary approach, and for state-of-the-art mathematical tools, which is able to forecast the spatial diffusion patterns of geographic events. Its time dynamics is quintessential for effective action. If an action based on a geographic analysis is a mistake, it is generally very costly both financially and may result in the loss of life. Classical ways to approach these issues have a differential-algebraic base, namely, making use of the already available data to calibrate a set of differential equations to provide the best possible prediction of geographic events' further unfolding. We can therefore speak of a differential-algebraic approach to the mathematical study of geographic events. The key to this modelling strategy is capturing the structure of the interactions among the main variables involved in the process, and to set the free parameters accordingly in order to reproduce the errors in the observable dynamic pattern as accurately as possible to obtain the best possible guarantee of future predictive accuracy. That is, the traditional geographic models, for example, for epidemics or predator-prey models of biomathematics, are mathematical models that rely on the correct encoding of the cause-effect relationships of the events in question (disease spread for example) and data is used to fit this model accordingly. Despite its intuitive appeal and it wide adoption, this approach has important limitations.

1. The interaction among variables needs to be modelled a priori, on the basis of previous data generated by the same process.
2. Large sets of temporal and spatial data are needed in order to effectively calibrate the model parameters.
3. The number of free parameters needed to attain a reasonable data fit could be relatively large.
4. Validation of the model is generally an interpolation based on data from the same epidemic, or an extrapolation based data from similar epidemics.
5. Even when the dynamical modelling provides useful cause-effect relations, calibrating the main parameters and estimating connection strengths between variables are difficult tasks. And parameters are typically changing in time as the phenomenon evolves.

A more systematic quantitative comparison between TWC and the other algorithms of geographic profiling may be found in, for example, [13]. We provide below a brief comparison between geographic profiling/geometric methods and topological methods of TWC in Sect. 6.3 after the chapter on the applications. The comparisons just use TWC—Alpha and TWC—Beta since the other TWC quantities have no correspondences with the classic geographic profiling methods.

References

1. P.L. Brantingham, P.J. Brantingham, *Environmental Criminology* (Waveland Press Inc, Prospect Heights IL, 1981)
2. D.K. Rossmo, *Geographic Profiling* (CRC Press, Boca Raton, FL, 2000), p.2000
3. N. Levine and Associates, *CrimeStat III—A Spacial Statistical Program for the Analysis of Crime Incident Locations* (The National Institute of Justice, Washington DC, 2004), pp.101–102
4. M. O'Leary, A new mathematical technique for geographic profiling, in *The NIJ Conference*, Washington DC, 17–19 June 2006
5. D. Canter, *Mapping Murder: The Secrets of Geographic Profiling* (Virgin Publishing, London, 2007)
6. S.C. Le Comber, D.K. Rossmo, A.N. Hassan, D.O. Fuller, J.C. Beier, Geographic profiling as a novel spatial tool for targeting infectious disease control. Int. J. Health Geograph. **10**, Num. 35 (8 pages) (2011)
7. S.C. Le Comber, B. Nicholls, D.K. Rossmo, P.A. Racey, Geographic profiling and animal foraging. J. Theor. Biol. **240**, 233–240 (2006)
8. P.M. Barone, R.M. Di Maggio, S. Mesturini, Materials for the study of the locus operandi in the search for missing persons in Italy, Forensic Sciences Research 2020 preprint, pp. 1–11 (published online February 23, 2021) (2020). https://doi.org/10.1080/20961790.2020.1854501

9. V. Berezowski, D. MacGregor, J. Ellis, I. Moffat, X. Mallett, More than an offender location tool: geographic profiling and body deposition sites. J. Police Crim. Psychol. (2021). https://doi.org/10.1007/s11896-021-09475-6
10. S. Curtis-Ham, W. Bernasco, O. N. Medvedev, D.L.L. Polasche, A national examination of the spatial extent and similarity of offenders' activity spaces using police data. ISPRS Int. J. Geo-Inf. **10**(2), 47 (2021). https://doi.org/10.3390/ijgi10020047
11. G. Hajela, M. Chawla, A. Rasool, A clustering based hotspot identification approach for crime prediction. Procedia Comput. Sci. **167**, 1462–1470 (2020)
12. D. Canter, T. Coffey, M. Huntley, C. Missen, Predicting serial killers' home base using a decision support system. J. Quant. Criminol. **16** Num. **4**, 457–478 (2000)
13. M. Buscema, E. Grossi, A. Bronstein, W. Lodwick, M.A. Zeydabadi, R. Benzi, F. Newman, A new algorithm for identifying possible epidemic sources with application to the German Escherichia Coli Outbreak. ISPRS Int. J. Geo-Inf. **2**, 155–200 (2013). https://doi.org/10.3390/ijgi2010155.

Chapter 3
The Theory of TWC

3.1 General Introduction to TWC Theory

TWC theory consists of several components. This chapter reviews the key elements that make up the theory and the constituent algorithms that are used in TWC. There are theoretical components of TWC, among them: (1) pseudo-distances, (2) attraction strength, (3) entropy, and (3) free energy, that are used in various TWC algorithms. The development of these components is followed by the theory of the TWC algorithms. The main goal of the topological approach is the extraction from the geographical data, of the dynamics and the patterns implicit in the data and make these explicit/visible. Figure 3.1 shows the main divisions of the TWC theory. Each division is further characterized by different strategic quantities and outputs.

The theoretical aspects of the five TWC algorithms depicted above can be summarized as follows.

1. TWC—Alpha represents a spatial estimate of the point or area where the process under examination "originated" (outbreak), where "originated" means where its impetus from an energy and entropy point of view. "Originated" in certain contexts means where the process started in other contexts means where the process would locate itself in the most efficient way with respect to energy expenditure. That is, the outbreak points can also be interpreted as the points from which the impetus of the process under examination is concentrated. This estimate is provided both at the point level (TWC alpha point) and at the "heat" map level (Alpha map).
2. TWC—Beta represents the current likely distribution of the process under consideration in the short term. The head map of Beta also represents the places where other hidden events of the same type could have occurred.
3. TWC—Gamma represents the likely future evolution of the TWC—Beta distribution, considering the system's self-organizing properties.

M. Buscema et al., *The Topological Weighted Centroid: A New Vision of Geographic Profiling*, Studies in Computational Intelligence 1095,
https://doi.org/10.1007/978-3-031-28901-9_3

4. TWC—Theta represents a further level of evolution over time, developed from TWC—Gamma as the "communication" and interaction between the observed events that stabilizes to become highly organized with the minimum energy required. TWC—Theta also provides a directed weighted graph from which a hypothesized flow of communication among the points is represented as connections.
5. TWC—Iota provides a "heat" map representing the infectivity rate of the area. It is also hypothesized as being a point where the process under consideration will terminate as will be described below. Iota can also represent where the beginning and end of the process associated with the given set of data point.

3.1.1 The Four Types of TWC

There are four types of TWC and each type will generate five types of maps. These four types are briefly described next. Chapter 5 will describe these four types in more detail. Three of the types (original, frequency, and semantic) deal with the entire input data set that TWC processes. The fourth, TWC-Windowing, is a process by which subsets, called windows, of the overall region can be processed and then "tiled" to focus phenomena on the subset of the entire region, the windows.

1. **TWC-Original**: The data occurrence at a geographical location is taken as one event regardless how many there are at the associated location. The phenomena is analyzed from the point of view of a single occurrence at each of the data point locations.
2. **TWC-Frequency**: The data at geographical locations take into account the number of occurrences of the phenomenon being analyzed.
3. **TWC-Semantic**: The data at geographical locations have, in addition to location data, latitude/longitude, attributes and then analyzed. An artificial adaptive system is augmented to the either the semantic TWC-Original or the semantic TWC-Frequency.
4. **TWC-Windowing**: Subregions of the entire area are the input data and then "tiled" together.

3.1.2 How the TWC Data Is Organized

TWC considers three types of data sets:

1. *Syntactic data*

$$Data = \{P\left(x_i, y_i\right)\}_{i=1}^{N}, \tag{3.1}$$

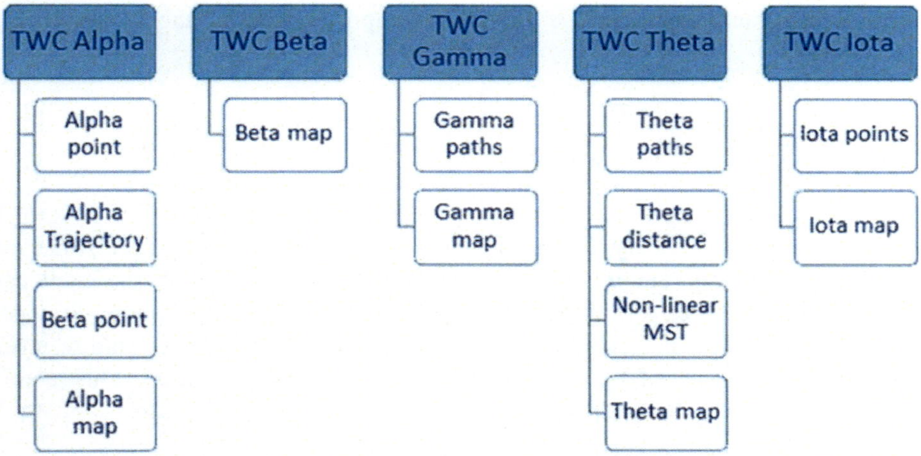

Fig. 3.1 TWC algorithms

where, N is the number of the data points ($N \gg 3$), and, x_i and y_i are longitude and latitude of the of ith point;

2. *Time series data*

$$Data = \left\{ \left\{ P\left(x_{i,t}, y_{i,t}\right) \right\}_{i=1}^{N} \right\}_{t=1}^{T}, \tag{3.2}$$

where Eq. (3.1) has to be modified to include a parameter that is associated with time, T is the number of the temporal data points and $x_{i,t}$ and $y_{i,t}$ are longitude and latitude of the point i at time step t;

3. *Semantic data*

$$Data = \left\{ \left\{ P\left(x_{i,t}, y_{i,t}, \left\{a_{i,k}\right\}_{k=1}^{M}\right) \right\}_{i=1}^{N} \right\}_{t=1}^{T}, \tag{3.3}$$

where each data-point is characterized by a set of specific features associated with Eq. (3.1), $a_{i,k}$ is the kth attribute associated with point i, and M is the number of the attributes (features) for each of the data points.

The Fig. 3.1 shows the main components of the TWC theory. Each algorithm is further characterized by different strategic quantities and outputs that are outlined next.

3.2 Theoretical Components of TWC

TWC input data points are assumed to have the following two characteristics:

1. Each point of the distribution represents a discrete manifestation of the same process;

2. The distribution of points is statistically representative of the process being analyzed.

3.2.1 Distance

The TWC approach is that the relationships among the data points are somehow related to "distances" among the points, where *distance* is not simply the Euclidean distance but distances related to energy or "attractions". These energy distances result in the subsequent distribution of the data points in the data set and result in various different types of TWC maps (alpha, beta, gamma, theta, iota). This section presents the TWC concept distance as it is used in the various algorithms.

The Concept of Pseudo-Distance and its Properties

This section introduces the concept called *pseudo-distance* (or equivalently, *indirect-distance*). In TWC, "distances" between two points i and j involve average distances between the two designated points and all other points. These two concepts are based on the traditional mathematical definition of distance.

Definition 3.1 A function $d : X \times X \rightarrow [0, \infty)$ is said to be a **distance** (in a metric space) if it satisfies the following properties.

$$1. d(x, y) \geq 0, \tag{3.4}$$

$$2. d(x, y) = 0 \text{ if and only if } x = y, \tag{3.5}$$

$$3. d(x, y) = d(y, x), \tag{3.6}$$

$$4. d(x, y) \leq d(x, z) + d(z, y). \tag{3.7}$$

Definition 3.2 The **Euclidean distance** between two points with two coordinates

$$P_i = (x_i, y_i), P_j = (x_j, y_j) \in \mathbb{R}^2$$

is

$$d(P_i, P_j) = d_{ij} = \sqrt{(x_i - x_j)^2 + (y_i - y_j)^2}. \tag{3.8}$$

Definition 3.3 Given data, that is, a subset of $N \geq 3$ points in the Euclidean space $\mathbb{R}^2$, the **pseudo-distance** between points P_i and P_j, with $i \neq j$, $P_i, P_j \in Dataset$, is defined to be the function

$$\overline{d}(\cdot, \cdot) : Data \times Data \rightarrow [0, \infty)$$

such that

$$\overline{d}(P_i, P_j) = \overline{d}(i, j) = \overline{d}_{ij} = \begin{cases} \frac{1}{N-2} \sum_{\substack{k \neq i \\ k \neq j}}^{N} d_{ik}, i \neq j \\ 0 \quad \text{for } i = j. \end{cases} \tag{3.9}$$

where d_{ij} is the Euclidean distance between points P_i and P_j. This concept can be generalized to $\mathbb{R}^n$ in a straight-forward manner, though we restrict ourselves to $\mathbb{R}^2$.

To calculate the pseudo-distance, we can also write it as follows.

$$\begin{aligned}\overline{d}_{ij} &= \frac{1}{N-2}\sum_{\substack{k\neq i\\k\neq j}}^{N} d_{ik}\\ &= \frac{1}{N-2}\left(\sum_{k=1}^{N} d_{ik} - d_{ij} - d_{ii}\right)\\ &= \frac{1}{N-2}\left(\sum_{k=1}^{N} d_{ik} - d_{ij}\right).\end{aligned}$$

The symbolic calculation of the pseudo-distance for $i = 2$ from a data set of 5 points is as follows.

$$\overline{d}_{21} = \frac{1}{5-2}\sum_{\substack{k\neq 2\\k\neq 1}}^{5} d_{2k} = \frac{1}{3}\left(d_{23} + d_{24} + d_{25}\right)$$

$$\overline{d}_{22} = 0$$

$$\overline{d}_{23} = \frac{1}{5-2}\sum_{\substack{k\neq 2\\k\neq 3}}^{5} d_{2k} = \frac{1}{3}\left(d_{21} + d_{24} + d_{25}\right)$$

$$\overline{d}_{24} = \frac{1}{5-2}\sum_{\substack{k\neq 2\\k\neq 4}}^{5} d_{2k} = \frac{1}{3}\left(d_{21} + d_{23} + d_{25}\right)$$

$$\overline{d}_{25} = \frac{1}{5-2}\sum_{\substack{k\neq 2\\k\neq 5}}^{5} d_{2k} = \frac{1}{3}\left(d_{21} + d_{23} + d_{24}\right)$$

Comment

The more a point is relatively close to most of the others, the less it contributes to the pseudo-distance. It can be defined as a system of distances because it takes into account not only the individual pairs but its relationship to all the points.

Moreover, given N points, this measure is a redistribution of the Euclidean distance among the N points. Euclidean distance is not a "systemic" measure in the following sense. If one of the given points changes its position, then only its distance from the other points changes but the distances among the other points do not change. The pseudo-distance, instead, is a systemic measure. If one point of the N given points changes its position, all the pseudo-distances among the other

$N - 1$ points are rearranged, in order to guarantee that the N points continue to form a "cluster". Generally speaking, the points that are more isolated from the others reduce their distance, while the points much closer to each other are pushed away. A pseudo-distance is not strictly speaking a "distance", because it does not adhere to all the axioms of a distance; for example, the distance from A to A is not zero, but it is the average of the distance from A from the other points. It is not symmetric because the distance from A to B is different of the distance from B to A. Moreover, the triangle inequality does not hold.

All the same the pseudo-distance presents some interesting features.

1. It is able to mimic the spatial behavior of biological systems such as groups of animals and also humans, which tend spontaneously to rearrange their reciprocal distances in order to maintain their group cohesion, especially in critical situations. If an individual member moves away from the group, all members reposition themselves accordingly in order to ensure that the group remains compact. This collective behavior is partially present in the plant world.
2. Given N points, the summation of all the differences (their "deltas") between the Euclidean distance and the pseudo-distance among these points is zero. Thus, the amount of energy between the distance matrices remains constant (conserves energy). That is, the matrix of the pseudo-distance is a weighted redistribution of the matrix of the Euclidean distance. The main "goal" of this redistribution is to maintain the cohesion of the given points as a cluster for the given points considered as a system.

Table 12 Points shows a possible distribution of N points ($N = 12$). Table ED, Table PD and Table $PDED$ below show respectively the Euclidean distance among them, the Pseudo-Distance and the matrix of their differences. This 12-point example is used in association with our development of the theoretical aspects of TWC. Later, two other examples of 10 points and 11 points are given in Chap. 4.

City	$Data$	$x - latitude$	$y - longitude$
$Volterra$	P_1	10.8662	43.3981
$Perugia$	P_2	12.3908	43.1076
$Cerveteri$	P_3	12.0974	41.9963
$Tarquinia$	P_4	11.7575	42.4235
$Populonia$	P_5	10.4917	42.9895
$Chiusi$	P_6	11.9448	43.0153
$Arezzo$	P_7	11.8792	43.4606
$Vetulonia$	P_8	10.9717	42.8572
$Bolsena$	P_9	11.9861	42.6438
$Roselle$	P_{10}	11.1385	42.808
$Veio$	P_{11}	12.4631	42.1121
$Vulci$	P_{12}	11.6311	42.4235

Table 12 Points

Fig. 3.2 Map of the distribution of 12 points

The term distance has the modifiers“pseudo” and “indirect” because, as we will see in Example 3.4, Eq. (3.9) may not satisfy two of the four properties necessary to be a distance. However, it is useful to understand the effect pseudo-distance produces on the points of space. As we mentioned above, the more a point is relatively close to most others, the less it contributes to the pseudo-distance. A point that is relatively close to many other points may therefore be close to another one in terms of pseudo-distance even if the two are physically distant in terms of Euclidean distance in the original space. Vice versa, a point that is physically close to another, but relatively distant from most others, may have a large pseudo-distant from the other despite being physically close to it in the original space. Therefore, application of the pseudo-distance expands or contracts the space between two points depending on their relative position in the global distribution. The pseudo-distance between two points is therefore based on the global distribution of points, and not on the individual property of the points as occurs in the Euclidean distance which only considers the two individual points, not the whole data set of points (Figs. 3.2, 3.3, 3.4 and 3.5).

Euclidean	P1	P2	P3	P4	P5	P6	P7	P8	P9	P10	P11	P12
P1	0.00	393.64	473.20	334.97	140.58	290.28	257.41	139.77	342.46	164.83	520.02	314.22
P2	393.64	0.00	291.51	236.44	482.59	115.51	157.65	365.48	156.12	326.58	253.15	259.29
P3	473.20	291.51	0.00	138.46	478.86	261.33	375.49	359.43	166.63	318.64	97.29	160.39
P4	334.97	236.44	138.46	0.00	351.67	157.43	264.84	227.64	80.52	184.82	195.61	32.06
P5	140.58	482.59	478.86	351.67	0.00	368.60	371.64	126.28	389.03	170.38	547.28	322.67
P6	290.28	115.51	261.33	157.43	368.60	0.00	114.16	250.04	94.80	211.15	264.11	169.88
P7	257.41	157.65	375.49	264.84	371.64	114.16	0.00	276.40	208.93	250.38	372.70	270.46
P8	139.77	365.48	359.43	227.64	126.28	250.04	276.40	0.00	262.91	44.11	422.84	200.17
P9	342.46	156.12	166.63	80.52	389.03	94.80	208.93	262.91	0.00	218.97	181.17	105.97
P10	164.83	326.58	318.64	184.82	170.38	211.15	250.38	44.11	218.97	0.00	379.50	158.49
P11	520.02	253.15	97.29	195.61	547.28	264.11	372.70	422.84	181.17	379.50	0.00	225.31
P12	314.22	259.29	160.39	32.06	322.67	169.88	270.46	200.17	105.97	158.49	225.31	0.00

Fig. 3.3 Table ED: the euclidean distance among the 12 points

Pseudo D	P1	P2	P3	P4	P5	P6	P7	P8	P9	P10	P11	P12
P1	0.00	297.77	289.82	303.64	323.08	308.11	311.40	323.16	302.89	320.65	285.14	305.71
P2	264.43	0.00	274.64	280.15	255.54	292.24	288.03	267.25	288.18	271.14	278.48	277.87
P3	264.80	282.97	0.00	298.28	264.24	285.99	274.57	276.18	295.46	280.26	302.39	296.08
P4	186.95	196.80	206.60	0.00	185.28	204.70	193.96	197.68	212.39	201.96	200.89	217.24
P5	360.90	326.70	327.07	339.79	0.00	338.10	337.80	362.33	336.06	357.92	320.23	342.69
P6	200.70	218.18	203.60	213.99	192.87	0.00	218.31	204.73	220.25	208.62	203.32	212.74
P7	266.26	276.24	254.46	265.52	254.84	280.59	0.00	264.37	271.11	266.97	254.73	264.96
P8	253.53	230.96	231.56	244.74	254.88	242.50	239.87	0.00	241.22	263.10	225.22	247.49
P9	186.50	205.14	204.09	212.70	181.85	211.27	199.86	194.46	0.00	198.85	202.63	210.15
P10	226.30	210.13	210.92	224.30	225.75	221.67	217.75	238.37	220.89	0.00	204.83	226.93
P11	293.90	320.58	336.17	326.34	291.17	319.49	308.63	303.62	327.78	307.95	0.00	323.37
P12	190.47	195.96	205.85	218.69	189.62	204.90	194.85	201.87	211.30	206.04	199.36	0.00

Fig. 3.4 Table PD: the pseudo-distance among the 12 points

Deltas	P1	P2	P3	P4	P5	P6	P7	P8	P9	P10	P11	P12	Tot
P1	0.00	95.86	183.38	31.33	-182.50	-17.83	-53.98	-183.39	39.57	-155.82	234.89	8.51	0.00
P2	129.20	0.00	16.87	-43.71	227.06	-176.73	-130.39	98.23	-132.07	55.44	-25.33	-18.58	0.00
P3	208.39	8.54	0.00	-159.82	214.62	-24.66	100.91	83.25	-128.83	38.38	-205.10	-135.69	0.00
P4	148.02	39.64	-68.14	0.00	166.39	-47.27	70.88	29.96	-131.87	-17.15	-5.27	-185.18	0.00
P5	-220.33	155.89	151.79	11.88	0.00	30.50	33.84	-236.05	52.97	-187.54	227.05	-20.02	0.00
P6	89.58	-102.66	57.73	-56.55	175.73	0.00	-104.16	45.31	-125.45	2.53	60.79	-42.86	0.00
P7	-8.85	-118.59	121.03	-0.68	116.80	-166.43	0.00	12.04	-62.18	-16.59	117.97	5.50	0.00
P8	-113.76	134.52	127.87	-17.10	-128.60	7.54	36.53	0.00	21.69	-218.99	197.62	-47.32	0.00
P9	155.95	-49.02	-37.46	-132.18	207.18	-116.47	9.07	68.45	0.00	20.12	-21.47	-104.19	0.00
P10	-61.47	116.45	107.72	-39.48	-55.36	-10.52	32.63	-194.27	-1.92	0.00	174.66	-68.44	0.00
P11	226.12	-67.43	-238.88	-130.73	256.11	-55.37	64.07	119.22	-146.62	71.55	0.00	-98.05	0.00
P12	123.75	63.32	-45.46	-186.63	133.05	-35.02	75.61	-1.70	-105.33	-47.55	25.95	0.00	0.00
Tot	676.62	276.53	376.44	-723.67	1130.48	-612.26	135.03	-158.94	-720.03	-455.63	781.76	-706.33	0.00

Fig. 3.5 Table PDED: difference among pseudo-distance and euclidean distance matrices

Property 3.1 *The pseudo-distance between i and j is generally different from that between j and i. In fact,*

$$\overline{d}_{ij} - \overline{d}_{ji} = \frac{1}{N-2}\left(\left(\sum_{k}^{N} d_{ik} - d_{ij}\right) - \left(\sum_{k}^{N} d_{jk} - d_{ji}\right)\right) \tag{3.10}$$

$$= \frac{1}{N-2}\left(\left(\sum_{k}^{N} d_{ik}\right) - \left(\sum_{k}^{N} d_{jk}\right)\right) + \frac{d_{ji} - d_{ij}}{N-2} \tag{3.11}$$

$$= \frac{1}{N-2}\left(\left(\sum_{k}^{N} d_{ik}\right) - \left(\sum_{k}^{N} d_{jk}\right)\right). \tag{3.12}$$

Since, in general,

$$\left(\sum_{k}^{N} d_{ik}\right) \neq \left(\sum_{k}^{N} d_{jk}\right),$$

one can see from (3.10) that the pseudo-distance is not always symmetric (violating the third property of distance, Eq. (3.6)).

Property 3.2 *The pseudo-distance may not satisfy the triangle inequality (3.7). In fact, using (3.10), we have*

$$\begin{aligned}\overline{d}_{ij} + \overline{d}_{ik} &= \frac{1}{N-2}\left(\left(\sum_{m}^{N} d_{im}\right) - d_{ij}\right) + \frac{1}{N-2}\left(\left(\sum_{m}^{N} d_{im}\right) - d_{ik}\right) \\ &= \frac{1}{N-2}\left(2\left(\sum_{m}^{N} d_{im}\right) - d_{ij} - d_{ik}\right), \end{aligned} \tag{3.13}$$

and

$$\overline{d}_{jk} = \frac{1}{N-2}\left(\sum_{m=1}^{N} d_{jm} - d_{jk}\right). \tag{3.14}$$

Thus, from Eqs. (3.13) and (3.9)

$$\begin{aligned}(\overline{d}_{ij} + \overline{d}_{ik}) - \overline{d}_{jk} &= \frac{1}{N-2}\left(2\left(\sum_{m}^{N} d_{im}\right) - d_{ij} - d_{ik}\right) - \frac{1}{N-2}\left(\sum_{m=1}^{N} d_{jm} - d_{jk}\right) \\ &= \frac{1}{N-2}\left(2\left(\sum_{m}^{N} d_{im}\right) - \sum_{m=1}^{N} d_{jm} - ((d_{ij} + d_{ik})) - d_{jk}\right). \end{aligned} \tag{3.15}$$

From (3.15), it can be seen that the condition for the triangle inequality can be violated. That is, the condition for the triangle inequality is

$$\begin{aligned}\overline{d}_{ij} + \overline{d}_{ik} &\geq \overline{d}_{jk} \\ (\overline{d}_{ij} + \overline{d}_{ik}) - \overline{d}_{jk} &\geq 0,\end{aligned}$$

which means that

$$\begin{aligned}\frac{1}{N-2}\left(2\left(\sum_{m}^{N} d_{im}\right) - \sum_{m=1}^{N} d_{jm} - ((d_{ij} + d_{ik}) - d_{jk})\right) &\geq 0 \\ 2\left(\sum_{m}^{N} d_{im}\right) - \sum_{m=1}^{N} d_{jm} &\geq (d_{ij} + d_{ik}) - d_{jk}. \end{aligned} \tag{3.16}$$

The right side of (3.16) is always positive since it is the Euclidean triangle inequality whereas the left side could be negative. An example where this occurs is given next.

Example 3.4 Suppose we have four points,

$$P_1 = (0, 1),\ P_2 = (a, 0),\ P_3 = (0, 2),\ \text{and } P_4 = (0, 0). \tag{3.17}$$

$$\begin{aligned}
\overline{d}_{12} + \overline{d}_{13} - \overline{d}_{23} &= \left(\frac{1}{2}(d_{13} + d_{14})\right) + \left(\frac{1}{2}(d_{12} + d_{14})\right) - \left(\frac{1}{2}(d_{21} + d_{24})\right) \\
&= \left(\frac{1}{2}d_{13} + d_{14}\right) - \frac{1}{2}d_{24} \\
&= \frac{3}{2} - \frac{a}{2}.
\end{aligned}$$

The triangle inequality states that distances have the property that $d_{12} + d_{13} \geq d_{23}$. However, if $a > 3$, for example, $a = 4$, the triangle inequality is violated. Moreover,

$$\begin{aligned}
\overline{d}_{24} - \overline{d}_{42} &= \frac{1}{2}(d_{21} + d_{23}) - \frac{1}{2}(d_{41} + d_{43}) \\
&= \frac{1}{2}\sqrt{a^2 + 1} + \frac{1}{2}\sqrt{a^2 + 4} - \frac{1}{2}(1 + 2).
\end{aligned}$$

The equation

$$\overline{d}_{24} - \overline{d}_{42} = 0,$$

if $a = 0$. Otherwise, $\overline{d}_{24} - \overline{d}_{42} > 0$.For example, if $a = 5$,

$$\overline{d}_{24} - \overline{d}_{42} = \frac{1}{2}\sqrt{26} + \frac{1}{2}\sqrt{29} - \frac{3}{2} = 3.7421... \neq 0.$$

Despite not being a proper distance, as we mentioned in our comments above, the pseudo-distance possesses interesting features from the viewpoint of our topological approach.

1. The pseudo-distance between two points depends on the distances of each point from all the others, which is a "holistic" characteristic of a data set of geographic entities.
2. Although pseudo-distance basically alters the notions of closeness and distance in the Euclidean sense, the sum of all the pseudo-distances across all points equals that of the classic Euclidean distances between the same points. This is a conservation property.

This latter characteristic is proved next.

Proposition 3.5 *Let M be the square matrix of distances of N points, that is*

$$(M)_{ij} = d_{ij},\ i, j = 1, ..., N,$$

and let $\overline{M}$ *be the matrix of pseudo-distances,* $\left(\overline{M}\right)_{ij} = \overline{d}_{ij}$*,and* $\Delta d_{ij} = d_{ij} - \overline{d}_{ij}$. *Then*

$$\sum_{i,j}^{N} \Delta d_{ij} = 0.$$

Proof Given $i \neq j$

$$\begin{aligned}\Delta d_{ij} &= d_{ij} - \overline{d}_{ij} \\ &= d_{ij} - \frac{1}{N-2}\left(\sum_{k}^{N} d_{ik} - d_{ij}\right) \\ &= \frac{N-1}{N-2} d_{ij} - \frac{1}{N-2}\sum_{k}^{N} d_{ik}.\end{aligned}$$

Since $\Delta d_{ii} = d_{ii} - \overline{d}_{ii} = 0$,

$$\begin{aligned}\sum_{j=1}^{N} \Delta d_{ij} &= \Delta d_{ii} + \sum_{j=1.j\neq i}^{N} \Delta d_{ij} \\ &= \sum_{j\neq i}^{N}\left(\frac{N-1}{N-2} d_{ij} - \frac{1}{N-2}\sum_{k}^{N} d_{ik}\right) \\ &= \frac{N-1}{N-2}\sum_{j\neq i}^{N} d_{ij} - \frac{1}{N-2}\sum_{j\neq i}^{N}\sum_{k}^{N} d_{ik} \\ &= \frac{N-1}{N-2}\sum_{j\neq i}^{N} d_{ij} - \frac{N-1}{N-2}\sum_{k}^{N} d_{ik} = 0.\end{aligned}$$

Thus,

$$\sum_{i=1}^{N}\sum_{j=1}^{N} \Delta d_{ij} = \sum_{i=1}^{N} 0 = 0$$

and the proposition holds. ■

We can calculate the summation over the rows and columns of the matrix Δd_{ij} as follows.

$$
\begin{aligned}
Sum(Row_i) &= \sum_{j=1}^{N} \Delta d_{ij} = \Delta d_{ii} + \sum_{j=1, j\neq i}^{N} \Delta d_{ij} \\
&= \frac{N-1}{N-2}\sum_{j=1}^{N} d_{ij} - \frac{1}{N-2}\sum_{j=1, j\neq i}^{N}\left(\sum_{k=1, j\neq i}^{N} d_{ik}\right) \\
&= \frac{N-1}{N-2}\sum_{j=1}^{N} d_{ij} - \frac{1}{N-2}\left((N-1)\sum_{j=1}^{N} d_{ij}\right) = 0.
\end{aligned} \tag{3.18}
$$

Likewise,

$$
\begin{aligned}
Sum(Col_j) &= \sum_{i=1}^{N} \Delta d_{ij} = \Delta d_{jj} + \sum_{i=1, i\neq j}^{N} \Delta d_{ij} \\
&= \frac{N-1}{N-2}\sum_{j=1}^{N} d_{ij} - \frac{1}{N-2}\sum_{i=1, i\neq j}^{N}\left(\sum_{k=1}^{N} d_{ik}\right),
\end{aligned} \tag{3.19}
$$

which is not in general zero, but as we have shown, the summation of the elements of the matrix is zero, that is,

$$
\sum_{i,j=1}^{N} \Delta d_{ij} = 0.
$$

Example 3.6 Consider the points given by data set (3.17).

$$
\begin{aligned}
&d_{11} = 0, d_{12} = d_{21} = \sqrt{a^2+1}, d_{13} = d_{31} = 1, d_{14} = d_{41} = 1, \\
&d_{21} = d_{12} = \sqrt{a^2+1}, d_{22} = 0, d_{23} = d_{32} = \sqrt{a^2+4}, d_{24} = d_{42} = a, \\
&d_{31} = d_{13} = 1, d_{32} = d_{32} = \sqrt{a^2+4}, d_{33} = 0, d_{34} = d_{43} = 2, \\
&d_{41} = d_{14} = 1, d_{42} = d_{42} = a, d_{43} = d_{43} = 2, d_{44} = 0,
\end{aligned}
$$

$$
\begin{aligned}
\sum_{ij} d_{ij} &= (\sqrt{a^2+1}+1+1) + (\sqrt{a^2+1}+\sqrt{a^2+4}+a) + (1+\sqrt{a^2}+4+2) + (1+a+2) \\
&= 2a + 2\sqrt{a^2+1} + 2\sqrt{a^2+4} + 8
\end{aligned}
$$

$$
D = \left[d_{ij}\right] = \begin{bmatrix} 0 & \sqrt{a^2+1} & 1 & 1 \\ \sqrt{a^2+1} & 0 & \sqrt{a^2+4} & a \\ 1 & \sqrt{a^2+4} & 0 & 2 \\ 1 & a & 2 & 0 \end{bmatrix}
$$

$$\overline{d}_{11} = 0, \overline{d}_{12} = \frac{1}{2}(d_{13} + d_{14}), \overline{d}_{13} = \frac{1}{2}(d_{12} + d_{14}), \overline{d}_{14} = \frac{1}{2}(d_{12} + d_{13})$$
$$\overline{d}_{11} = 0, \overline{d}_{12} = \frac{1}{2}(1 + 1), \overline{d}_{13} = \frac{1}{2}\left(\sqrt{a^2+1} + 1\right), \overline{d}_{14} = \frac{1}{2}\left(\sqrt{a^2+1} + 1\right)$$

$$\overline{d}_{21} = \frac{1}{2}(d_{23} + d_{24}), \overline{d}_{22} = 0, \overline{d}_{23} = \frac{1}{2}(d_{21} + d_{24}), \overline{d}_{24} = \frac{1}{2}(d_{21} + d_{23})$$
$$\overline{d}_{21} = \frac{1}{2}\left(\sqrt{a^2+4} + a\right), \overline{d}_{22} = 0, \overline{d}_{23} = \frac{1}{2}\left(\sqrt{a^2+1} + a\right), \overline{d}_{24} = \frac{1}{2}\left(\sqrt{a^2+1} + \sqrt{a^2+4}\right)$$

$$\overline{d}_{31} = \frac{1}{2}(d_{32} + d_{34}), \overline{d}_{32} = \frac{1}{2}(d_{31} + d_{34}), \overline{d}_{33} = 0, , \overline{d}_{34} = \frac{1}{2}(d_{31} + d_{32})$$
$$\overline{d}_{31} = \frac{1}{2}\left(\sqrt{a^2+4} + 2\right), \overline{d}_{32} = \frac{1}{2}(1 + 2), \overline{d}_{33} = 0, \overline{d}_{34} = \frac{1}{2}\left(1 + \sqrt{a^2+4}\right)$$

$$\overline{d}_{41} = \frac{1}{2}(d_{42} + d_{43}), \overline{d}_{42} = \frac{1}{2}(d_{41} + d_{43}), \overline{d}_{43} = \frac{1}{2}(d_{41} + d_{42}), \overline{d}_{44} = 0$$
$$\overline{d}_{41} = \frac{1}{2}(a + 2), \overline{d}_{42} = \frac{1}{2}(1 + 2), \overline{d}_{43} = \frac{1}{2}(1 + a), \overline{d}_{44} = 0$$

$$\begin{aligned}\sum_{ij} \overline{d}_{ij} &= \frac{1}{2}(1+1) + \frac{1}{2}\left(\sqrt{a^2+1} + 1\right) + \frac{1}{2}\left(\sqrt{a^2+1} + 1\right) + \\ &\frac{1}{2}\left(\sqrt{a^2+4} + a\right) + \frac{1}{2}\left(\sqrt{a^2+1} + a\right), + \frac{1}{2}\left(\sqrt{a^2+1} + \sqrt{a^2+4}\right) + \\ &\frac{1}{2}\left(\sqrt{a^2+4} + 2\right) + \frac{1}{2}(1+2) + \frac{1}{2}\left(1 + \sqrt{a^2+4}\right) + \\ &\frac{1}{2}(a+2) + \frac{1}{2}(1+2) + \frac{1}{2}(1+a+2) \\ &= 2a + 2\sqrt{a^2+1} + 2\sqrt{a^2+4} + \frac{17}{2}\, 2a + 2\sqrt{a^2+1} + 2\sqrt{a^2+4} + 8\end{aligned}$$

$$\overline{D} = \left[\overline{d}_{ij}\right] = \begin{bmatrix} 0 & 1 & \frac{1}{2} + \frac{\sqrt{a^2+1}}{2} & \frac{1}{2} + \frac{\sqrt{a^2+1}}{2} \\ \frac{\sqrt{a^2+4}}{2} + \frac{a}{2} & 0 & \frac{\sqrt{a^2+1}}{2} + \frac{a}{2} & \frac{\sqrt{a^2+1}}{2} + \frac{\sqrt{a^2+4}}{2} \\ \frac{\sqrt{a^2+4}}{2} + 1 & \frac{3}{2} & 0 & \frac{\sqrt{a^2+4}}{2} + \frac{1}{2} \\ \frac{a}{2} + 1 & \frac{3}{2} & \frac{a}{2} + \frac{3}{2} & 0 \end{bmatrix}.$$

Example 3.7 (Distance and Pseudo-distance Matrices) Suppose we are given a the table of 10 points below.

Data	x	y
P_1	2	11
P_2	2	13
P_3	35	14
P_4	28	14
P_5	30	15
P_6	34	10
P_7	35	12
P_8	33	16
P_9	31	10
P_{10}	28	12

Table of 10 Points

The distance matrix for these 10 points is as follows.

d_{ij}	$j=1$	$j=2$	$j=3$	$j=4$	$j=5$	$j=6$	$j=7$	$j=8$	$j=9$	$j=10$
$i=1$	0.00	2.00	33.14	26.17	28.28	32.02	33.02	31.40	29.02	26.02
$i=2$	2.00	0.00	33.02	26.02	28.07	32.14	33.02	31.14	29.15	26.02
$i=3$	33.14	33.02	0.00	7.00	5.10	4.12	2.00	2.83	5.66	7.28
$i=4$	26.17	26.02	7.00	0.00	2.24	7.21	7.28	5.39	5.00	2.00
$i=5$	28.28	28.07	5.10	2.24	0.00	6.40	5.83	3.16	5.10	3.61
$i=6$	32.02	32.14	4.12	7.21	6.40	0.00	2.24	6.08	3.00	6.32
$i=7$	33.02	33.02	2.00	7.28	5.83	2.24	0.00	4.47	4.47	7.00
$i=8$	31.40	31.14	2.83	5.39	3.16	6.08	4.47	0.00	6.32	6.40
$i=9$	29.02	29.15	5.66	5.00	5.10	3.00	4.47	6.32	0.00	3.61
$i=10$	26.02	26.02	7.28	2.00	3.61	6.32	7.00	6.40	3.61	0.00

The pseudo-distance matrix for these 10 points is as follows.

$\overline{d}_{ij}$	$j=1$	$j=2$	$j=3$	$j=4$	$j=5$	$j=6$	$j=7$	$j=8$	$j=9$	$j=10$
$i=1$	0.00	29.88	25.99	26.86	26.60	26.13	26.01	26.21	26.51	26.88
$i=2$	29.82	0.00	25.95	26.82	26.56	26.05	25.95	26.18	26.43	26.82
$i=3$	8.38	8.39	0.00	11.64	11.88	12.00	12.27	12.16	11.81	11.61
$i=4$	7.77	7.79	10.16	0.00	10.76	10.14	10.13	10.36	10.41	10.79
$i=5$	7.44	7.47	10.34	10.69	0.00	10.17	10.25	10.58	10.34	10.52
$i=6$	8.44	8.42	11.93	11.54	11.64	0.00	12.16	11.68	12.07	11.65
$i=7$	8.29	8.29	12.17	11.51	11.69	12.14	0.00	11.86	11.86	11.54
$i=8$	8.23	8.26	11.80	11.48	11.76	11.39	11.59	0.00	11.36	11.35
$i=9$	7.79	7.77	10.71	10.79	10.78	11.04	10.86	10.63	0.00	10.97
$i=10$	7.78	7.78	10.12	10.78	10.58	10.24	10.16	10.23	10.58	0.00

The next table has the differences between distances and pseudo-distance matrices $\Delta M_{N\times N} = M_{N\times N} - \overline{M}_{N\times N}$.

$\Delta M_{N\times N}$	$j=1$	$j=2$	$j=3$	$j=4$	$j=5$	$j=6$	$j=7$	$j=8$	$j=9$	$j=10$	sum
$i=1$	0.00	−27.88	7.15	−0.69	1.69	5.88	7.01	5.19	2.51	−0.86	0
$i=2$	−27.82	0.00	7.07	−0.80	1.51	6.09	7.07	4.97	2.73	−0.80	0
$i=3$	24.76	24.62	0.00	−4.64	−6.78	−7.88	−10.27	−9.34	−6.15	−4.33	0
$i=4$	18.41	18.23	−3.16	0.00	−8.52	−2.93	−2.85	−4.98	−5.41	−8.79	0
$i=5$	20.85	20.61	−5.24	−8.46	0.00	−3.77	−4.41	−7.42	−5.24	−6.92	0
$i=6$	23.58	23.72	−7.80	−4.33	−5.24	0.00	−9.93	−5.60	−9.07	−5.33	0
$i=7$	24.73	24.73	−10.17	−4.23	−5.86	−9.90	0.00	−7.38	−7.38	−4.54	0
$i=8$	23.18	22.89	−8.97	−6.09	−8.59	−5.31	−7.12	0.00	−5.04	−4.95	0
$i=9$	21.23	21.38	−5.05	−5.79	−5.68	−8.04	−6.39	−4.30	0.00	−7.36	0
$i=10$	18.24	18.24	−2.84	−8.78	−6.98	−3.92	−3.16	−3.83	−6.98	0.00	0

Remark 3.8 The negative sign in an entry of the difference matrix shows the points for which the space is "expanding", while the points with positive entries show the points for which the space is "contracting". Moreover, when the pseudo-distance matrix is treated as a distance matrix, and the pseudo-distance matrix of this distance matrix is calculated, that is the meta-distance, the result defines a global attractor distribution. Thus, the meta-distance is considered as being an attractor.

3.2.2 Attraction Strength

What is termed in TWC as *attraction strength* is an exponential that decays as the distance (pseudo-distance) becomes larger. It quantifies the attraction strength of each given point with respect to all others, where strengths exponentially decay with indirect distance. Historically, the concept of strength of attraction has always been linked to objects with a mass. This section proposes a definition that extends this concept to *objects* (occurrences of a disease, a robbery, and so on) with no associated mass. For TWC we define attraction as linked with the concept of pseudo-distances.

Definition 3.9 (*Attraction Strength*) Given N points in a 2-dimensional space, the **non-parametric attraction strength** w_i that a point i exerts on all the others is given by

$$w_i = \frac{1}{N-1} \sum_{j \neq i}^{N} e^{-\frac{\overline{d}_{ij}}{D}}$$

where $\overline{d}_{ij}$ is the pseudo-distance between points i and j and $D = \max_{i,j} \sum \left\{\overline{d}_{ij}\right\}$. The **parametric (with parameter α_n) strength of attraction** is given by

$$w_i(\alpha_n) = \frac{1}{N-1} \sum_{j=1, j \neq i}^{N} e^{-\frac{\overline{d}_{ij}}{D}\alpha_n} \tag{3.20}$$

and $\alpha_n \geq 0$ corresponds to the algorithm being used, and it is the inverse of the simulated temperature of the entire system of points at any step. For TWC Semantic

$$w_i\left(\alpha_n\right) = \frac{1}{N-1} \sum_{j \neq i}^{N} e^{-\frac{\overline{d}_{ij} + d_{ij}^{[CM]}}{D}\alpha_n}$$

where the α_n is a natural number and modulates the attraction strength of points over time according to the rule $\alpha_{n+1} = \alpha_n + \in$ and $d_{ij}^{[CM]}$ arise from the contractive map algorithm, ACM (see [1, 2]). A self-attraction **strength** is an attraction strength where the point itself is not excluded from the attraction and in this case there are N points, not $N - 1$. This is denoted by the parameter as β ($\beta_{n+1} = \beta_n + \in$) so that

$$w_i\left(\beta_n\right) = \frac{1}{N} \sum_{j=1}^{N} e^{-\frac{\overline{d}_{ij}}{D}\beta_n}$$

and the Semantic TWC for the self-attraction case is,

$$Sw_i\left(\beta_n\right) = \frac{1}{N} \sum_{j=1}^{N} e^{-\frac{\overline{d}_{ij} + d_{ij}^{[CM]}}{D}\beta_n}.$$

Remark 3.10 The stronger the attraction strength is at one point, the more such a point "communicates" with the others, that is, the more its location becomes "meaningful" once put in relation to all the other points. The non-parametric version is used to calculate the "absolute" strength value of each point. The parametric version is used in the calculation of TWC alpha trajectories (see below). The "self" version is used to identify the beta star parameter. That is, Eq. (3.20) quantifies the attraction strength of each given point with respect to all others, where strengths exponentially decay with respect to the pseudo-distance. If each point of the data set has an associated energy in the plane, then the stronger the attraction strength at one point has, the more such a point influences each of the other points.

3.2.3 TWC Trajectories

What TWC terms as the (discrete) **trajectory** is the path connecting the center of mass with "optimal" points calculated by the five TWC algorithms, alpha, beta, gamma, or iota. Given the set of points (x, y) from the grid that, once drawn on a Cartesian plane, determine a trajectory whose starting point is the center of mass of the assigned (given data) points and the arrival point is called TWC alpha or beta.

$$TWC_x(\alpha_n) = \frac{1}{\sum_{i=1}^{N} w_i(\alpha_n)} \sum_{i=1}^{N} w_i(\alpha_n) x_i, \tag{3.21}$$

$$TWC_y(\alpha_n) = \frac{1}{\sum_{i=1}^{N} w_i(\alpha_n)} \sum_{i=1}^{N} w_i(\alpha_n) y_i. \tag{3.22}$$

forms a trajectory. That is, the trajectory is constructed by weighing the coordinates of the center of mass with the attraction strength of the individual points as the alpha parameter changes. The alpha parameter simulates, as in thermodynamic systems, the inverse of the temperature.

Remark 3.11 The alpha (beta, gamma, theta, iota) path is the vector of points starting from the center of mass and arriving to the alpha point. The more the alpha path is nonlinear, the more complex is the set of assigned (input data) points, their distribution. In this case it means that the assigned points interact strongly generating many side effects. That is, any minimal displacement of some of assigned points may generate a strong rearrangement of all the others. Each point of the alpha path that is closer to the alpha point represents a place where there is a reduction of the entropy of the distances among the assigned points. The Euclidean distance between any component of the alpha path vector and the next component represents the instant "velocity" of an abstract object in that position along the path. The alpha path, in short, should represent the ideal way that is the safest path from the center of mass to the alpha point of a prey that tries to control, in the best way, the "attacks" coming from the assigned points. Figure 3.6 shows the alpha path for the 12 point example.

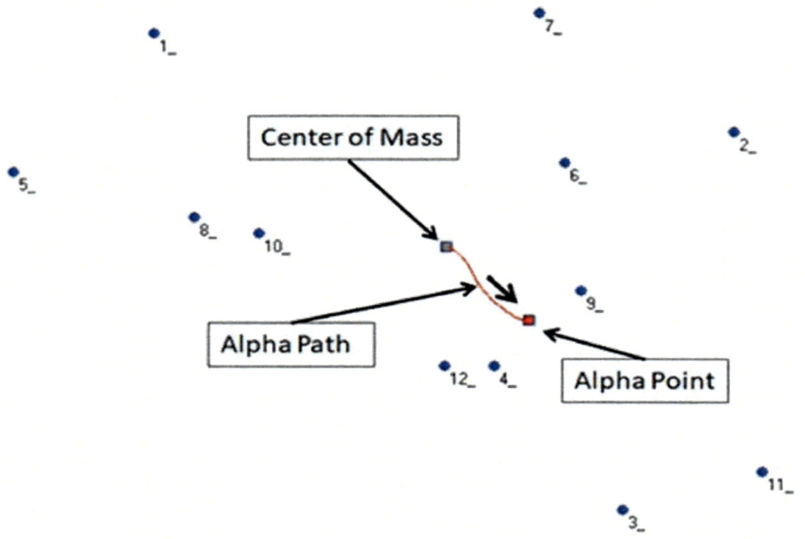

Fig. 3.6 The alpha path from center of mass and alpha point

3.2.4 Entropy

The information theory concept of entropy is redefined in the context of a distribution of points. Entropy measures the organization of a system. However, *information* in the context of geographical assigned points needs to be defined first.

Definition 3.12 (*Information*) The **information** contained in an event E emitted from a source, is calculated as

$$I\left(\mathcal{E}\right) = -\log_b \mathbb{P}\left(\mathcal{E}\right), \tag{3.23}$$

where b is the base of the logarithm and $\mathbb{P}\left(\mathcal{E}\right)$ is the probability of event $\mathcal{E}$. Usually $b = 2$.

Remark 3.13 The main idea behind Definition 3.12 is that the less likely an event is, the more information it carries when it occurs. Moreover, the place from which our information will emanate is our topological weighted centroid.

A natural consequence, once the information of an event has been defined, is that it is possible to define the entropy of the system.

Definition 3.14 (*Shannon Entropy*) The Shannon **entropy** of a source X is defined as the expected value of the information,

$$\mathbb{H}\left(X\right) = \mathbb{E}(-\log_b \mathbb{P}\left(X\right)) = -\sum_{i=1}^{N} \mathbb{P}_i\left(X\right) \log_b \mathbb{P}_i\left(X\right). \tag{3.24}$$

The last equality of Eq. (3.24) holds if X is a discrete random variable. The objective at this point is to find a probability associated with the points of the starting data set so we can define an analogous measure of entropy for our case. To do so, the weights (attraction strength), $w_i(\alpha_n)$, need to be normalized so that it is akin to a probability,

$$p_i(\alpha_n) = \frac{w_i(\alpha_n)}{\sum_{i=1}^{N} w_i(\alpha_n)}.$$

For TWC, entropy H is defined as follows.

Definition 3.15 (*TWC Entropy*) The **TWC entropy** for each iteration of TWC(n), that is, for each n, is defined, according to the Shannon entropy, as

$$H(\alpha_n) = -\sum_{i=1}^{N} p_i(\alpha_n) \log_2 (p_i(\alpha_n)), \tag{3.25}$$

where α_n is our parameter.

Remark 3.16 The value of the entropy at the center of mass is $H(\alpha_0) = H(0) = log_2 N$, so that the entropy is maximum. The value of the entropy decreases monotonically from the center of mass until it attains a minimal value when $H(\alpha_{n\to\infty}) = H(\alpha_n \to \infty)$. This is the TWC point that corresponds to where the summation of $\overline{d}_{ij}$ is minimal. This implies that, with each iteration, it will be possible to calculate the probability of each $w_i(\alpha_n)$ and consequently the information of each one by means of Eq. (3.23).

3.2.5 *Free Energy*

This section introduces the concept of "free energy" from physics. However, we adapt this concept to the case of a distribution of points. In physics, free energy, F, is obtained from the internal energy, U, minus the product of temperature and entropy, TS,

$$F = U - TS. \tag{3.26}$$

Free energy is a measure of the amount of energy available to do work. Whenever an energy change takes place, the total amount of energy stays the same, in a conservative system. As the entropy increases, the disorder energy increases and free energy decreases. The concepts and laws coming from physics, in this context, free energy and entropy, are useful concepts for understanding our approach. We use the concept of a thermodynamical system as function of T, where $T = 1/\alpha_n$, α_n our parameter, to calculate the free energy in the context of a distribution of points.

TWC uses the concept of free energy in the context of geometric distributions of events in a time/space geographic domain that is akin to concept of free energy in thermodynamics. This is done as follows.

Definition 3.17 Given

$$Z(\alpha_n) = \sum_{i=1}^{N} w_i(\alpha_n),$$

the free energy of the system is

$$F(\alpha_n) = -\frac{\ln Z(\alpha_n)}{\alpha_n}. \tag{3.27}$$

Remark 3.18 The parameter alpha plays the role of the inverse of temperature. The singular points of free energy at the high temperature phase, where

$$T = \frac{1}{\alpha_n}$$

is close to the center of mass (is close to the low temperature phase, or is close to the maximum density), and are separated by a region where the free energy is maximum at least at one point. The more a point is relatively close to most of the others, the less it contributes to the distance. It can be defined as a system distance because it takes into account not only the individual pairs but the set of all the points (this insight is due to Roberto Benzi).

3.2.6 *Alpha Star*

There is a (unique) point α^* of maximum free energy,

$$F(\alpha^*) = \max_{\alpha_n} \{F(\alpha_n)\}. \tag{3.28}$$

Theorem 3.19 $\max_{\alpha_n} \{F(\alpha_n)\}$ *exists and is unique, that is,* α^**, such that* $F(\alpha^*) = \max_{\alpha_n} \{F(\alpha_n)\}$ *is unique.*

Proof Change the notation of $F(\alpha_n) = -\frac{\ln Z(\alpha_n)}{\alpha_n}$ to $F(x) = -\frac{\ln Z(x)}{x}$. Then

$$
\begin{aligned}
F(x) &= -\frac{\ln Z(x)}{x} \\
&= -\frac{\ln \sum w_i(x)}{x} \\
&= -\frac{\ln\left(\frac{1}{N-1}\sum_{j\neq i}^{N} e^{-\frac{\overline{d}_{ij}}{D}x}\right)}{x}.
\end{aligned}
$$

Now, $x > 0$ so that $F(x)$ is continuous. By construction, the sequence α_n, our x, is an increasing sequence that has an upper limit. Thus $x \in [\alpha_0, A]$, $\alpha_0 > 0$. This means that $F(x)$ has a maximum in $[\alpha_0, A]$ (since F is continuous). Thus, an optimal exists.

Next let us look at uniqueness.

$$
F'(x) = -\left(\frac{1}{\sum w_i(x)}\left(\sum w_i'(x)\right)x^{-1} - \left(\ln \sum w_i(x)\right)x^{-2}\right).
$$

But

$$
w_i'(x) = -\frac{\overline{d}_{ij}}{D} w_i(x),
$$

so that

$$
\begin{aligned}
F'(x) &= -\left(\frac{1}{\sum w_i(x)}\left(\sum -\frac{\overline{d}_{ij}}{D} w_i(x)\right)x^{-1} - \left(\ln \sum w_i(x)\right)x^{-2}\right) \\
&= \frac{1}{\sum w_i(x)}\left(\sum \frac{\overline{d}_{ij}}{D} w_i(x)\right)x^{-1} + \left(\ln \sum w_i(x)\right)x^{-2}
\end{aligned}
$$

This means that $F'(x) = 0$ for

$$
\begin{aligned}
\frac{1}{\sum w_i(x)}\left(\sum \frac{\overline{d}_{ij}}{D} w_i(x)\right)x^{-1} &= -\left(\ln \sum w_i(x)\right)x^{-2} \\
\frac{1}{\sum w_i(x)}\left(\sum \frac{\overline{d}_{ij}}{D} w_i(x)\right) &= F(x).
\end{aligned}
$$

This is a fixed point problem. A necessary condition is that $\left|F'(x)\right| \leq 1$. This condition means that

$$\left|\frac{1}{\sum w_i(x)}\left(\sum \frac{\overline{d}_{ij}}{D} w_i(x)\right) x^{-1} + \left(\ln \sum w_i(x)\right) x^{-2}\right| \leq 1$$

$$\left|\frac{1}{\sum w_i(x)}\left(\sum \frac{\overline{d}_{ij}}{D} w_i(x)\right) + \left(\ln \sum w_i(x)\right) x^{-1}\right| \leq x$$

$$\left|\frac{1}{\sum w_i(x)}\left(\sum \frac{\overline{d}_{ij}}{D} w_i(x)\right) - F(x)\right| \leq x$$

so that

$$-x \leq \frac{1}{\sum w_i(x)}\left(\sum \frac{\overline{d}_{ij}}{D} w_i(x)\right) - F(x) \leq x$$

is our fixed point condition. Now

$$\frac{\min \overline{d}_{ij}}{D} \leq \frac{1}{\sum w_i(x)}\left(\sum \frac{\overline{d}_{ij}}{D} w_i(x)\right) \leq 1$$

so that

$$\frac{\min \overline{d}_{ij}}{D} - F(x) \leq \frac{1}{\sum w_i(x)}\left(\sum \frac{\overline{d}_{ij}}{D} w_i(x)\right) - F(x) \leq 1 - F(x).$$

The condition then is that

$$\begin{aligned} -x &\leq \frac{\min \overline{d}_{ij}}{D} - F(x) \leq 1 - F(x) \leq x, \\ -x + F(x) &\leq \frac{\min \overline{d}_{ij}}{D} \leq 1 \leq x + F(x). \end{aligned} \tag{3.29}$$

This expression is dominated by x and x increases so that Eq. (3.29) holds. In fact it holds with strict inequality so that the fixed point (or maximum) is unique. ■

The maximum value of free energy iteration therefore represents a unique point. After the point associated with α^*, the entropy and free energy begins to decrease together, while the temperature monotonically decreases.

3.3 TWC Algorithms

The five TWC algorithms depict, in the form of a map, what we call a *scalar field*. In general, the scalar field or TWC map is as follows.

Definition 3.20 (*Scalar Field*) A **scalar field** or **TWC map** defined on a region of space U is a function $s : U \rightarrow X$, where $X \in \mathbb{R}$ (a real number) that associates each point of $U \subset \mathbb{R}^n$ (in our case, a point is (x, y), $\mathbb{R}^2$, $2-D$) with a scalar value, a real number. The real number will be different for each of our five TWC types.

The five types of TWC algorithms use the components discussed in the previous sections. However, each type has its own approach to the implementation of these components. TWC theory is based on the definition of a different type of "centroid". This point is no longer only a function of the Euclidean distances of the points but also depends on their pseudo-distances and is modulated by a control parameter α_n (β, γ). By considering distances in the modified space, that is, the space in which pseudo-distances have modified the relationships between points by expanding or contracting the Euclidean distance, it is possible to take into account the structure of the distribution of points as a whole and in particular its varying local levels of organization, which represent, from the TWC point of view, a key piece of information used inferring the temporal unfolding of the process represented by the way the points geographically are found. Topological weighted centroid is defined next where the parameter α_n could also be β, γ.

Definition 3.21 (*Topological Weighted Centroid*) Given $\mathfrak{D} = \left\{(x_i, y_i)_{i-1}^N\right\} \subset \mathbb{R}^2$, the **topological weighted centroid** (TWC) of the points belonging to $\mathfrak{D}$ for any specific value of the control parameter α_n is given by the point $TWC(\alpha_n) = \left(TWC_x(\alpha_n), TWC_y(\alpha_n)\right) \in \mathbb{R}^2$, where the coordinates are:

$$TWC_x(\alpha_n) = \frac{\sum_{i=1}^{N} w_i(\alpha_n)x_i}{\sum_{i=1}^{N} w_i(\alpha_n)}, \tag{3.30}$$

$$TWC_y(\alpha_n) = \frac{\sum_{i=1}^{N} w_i(\alpha_n)y_i}{\sum_{i=1}^{N} w_i(\alpha_n)}. \tag{3.31}$$

Remark 3.22 For the TWC algorithms, $\alpha_0 = 0$, $\alpha_{n+1} = \alpha_n + 0.1$ (α being a generic parameter and could be β, γ) where these values are used in the implementation of all the algorithms.

Proposition 3.23 *Given the $TWC(\alpha_n)$ for a set $\mathfrak{D}$ of N points in $\mathbb{R}^2$, then,*

$$\begin{aligned} TWC(\alpha_0) &= \left(\tfrac{1}{N}\sum_{i=1}^{N} x_i, \tfrac{1}{N}\sum_{i=1}^{N} y_i\right) \\ \lim_{n\rightarrow\infty} TWC(\alpha_n) &= \tfrac{1}{2R}\left(\sum_{k=1}^{R}\left(x_{ik}, x_{jk}\right), \sum_{k=1}^{R}\left(y_{ik}, y_{jk}\right)\right), \end{aligned} \tag{3.32}$$

where $(x_{ik}, x_{jk}), (y_{ik}, y_{jk}) = \arg\min T_{ij}, i, j \in \{1, ..., N\}$ *identify the points between which the pseudo-distance is minimal. There could be more than one such minimal pseudo-distance.*

Proof (a) If $n = 0, \alpha_0 = 0,$ *and*

$$TWC_x(0) = \frac{\sum_{i=1}^{N} w_i(0)x_i}{\sum_{i=1}^{N} w_i(0)} = \frac{\sum_{i=1}^{N}\frac{1}{N-1}\sum_{j\neq i,j=1}^{N} e^{-\frac{\bar{d}_{ij}}{D}0}x_i}{\sum_{i=1}^{N}\frac{1}{N-1}\sum_{j\neq i,j=1}^{N} e^{-\frac{\bar{d}_{ij}}{D}0}}$$

$$= \frac{\frac{1}{N-1}\sum_{i=1}^{N}x_i\sum_{j\neq i,j=1}^{N}1}{\frac{1}{N-1}\sum_{i=1}^{N}\sum_{j\neq i,j=1}^{N}1}$$

$$= \frac{\frac{1}{N-1}\sum_{i=1}^{N}x_i}{\frac{N}{N-1}} = \frac{1}{N}\sum_{i=1}^{N}x_i.$$

Similarly,

$$TWC_y(0) = \frac{1}{N}\sum_{i=1}^{N}y_i.$$

This is the center of mass. That is, the algorithm starts at the center of mass.

(b) For TWC, $\alpha_n, n > 0$, the case of the x coordinate will be proved. The y coordinate is done in the same way.

$$TWC_x(\alpha_n) = \frac{\sum_{i=1}^{N} w_i(\alpha_n)x_i}{\sum_{i=1}^{N} w_i(\alpha_n)} = \frac{\sum_{i=1}^{N}\frac{1}{N-1}\sum_{j\neq i,j=1}^{N} e^{-\frac{\bar{d}_{ij}}{D}\alpha_n}x_i}{\sum_{i=1}^{N}\frac{1}{N-1}\sum_{j\neq i,j=1}^{N} e^{-\frac{\bar{d}_{ij}}{D}\alpha_n}}.$$

Since the free energy identifies the optimal α, at the optimal alpha, α^*, we have

$$TWC_x(\alpha^*) = \frac{\sum_{i=1}^{N} w_i(\alpha^*) x_i}{\sum_{i=1}^{N} w_i(\alpha^*)} = \frac{\sum_{i=1}^{N} \frac{1}{N-1} \sum_{j \neq i, j=1}^{N} e^{-\frac{\bar{d}_{ij}}{D}\alpha^*} x_i}{\sum_{i=1}^{N} \frac{1}{N-1} \sum_{j \neq i, j=1}^{N} e^{-\frac{\bar{d}_{ij}}{D}\alpha^*}},$$

which is the second the above equation. ∎

The TWC weighted centroid at the zeroth iteration coincides with the classic center of mass

$$CM = \left(\frac{1}{N} \sum_{i=1}^{N} x_i, \frac{1}{N} \sum_{i=1}^{N} y_i \right)$$

whose coordinates are nothing more than the average of the coordinates of all the points of the given set. As α_n increases monotonically with a small quantity $\alpha_{n+1} = \alpha_n + \epsilon$, we use $\frac{1}{10}$ as mentioned above, the centroid moves away from the classic center of mass. Definition 3.21 considers a centroid point for each iteration of the algorithm, that is, for each element of the alpha sequence. This means that, considering very small values of $\epsilon > 0$, when α_n varies, a trajectory will be drawn. Since what we are looking for is only one point that can constitute the centroid not merely topologically but weighed by the intrinsic semantics of the points of $\mathfrak{D}$, it is necessary to determine the optimal alpha value. In order to do this, some objective functions will be defined in the following sections. To this end, the two concepts previously defined, entropy and free energy, in the context of a geographic distribution of points, are used.

3.4 TWC-Alpha

The computation of the TWC-Alpha map uses the concept of alpha star. Recall that

$$F(\alpha^*) = \max_{\alpha_n} \{F(\alpha_n)\}.$$

It is this alpha star that is used.

3.4.1 TWC Alpha Point

$$TWC_x(\alpha^*) = \frac{1}{\sum_{i=1}^{N} w_i(\alpha_n)} \sum_{i=1}^{N} w_i(\alpha^*) x_i, \tag{3.33}$$

$$TWC_y(\alpha^*) = \frac{1}{\sum_{i=1}^{N} w_i(\alpha_n)} \sum_{i=1}^{N} w_i(\alpha^*) y_i. \tag{3.34}$$

Remark 3.24 The TWC alpha point is the point where the free energy is maximum and, as demonstrated experimentally, the entropy of the system is minimal. It is therefore a point of maximum organization. For this reason, it was elected as the point of the possible origin (impetus) of the system's dynamics. In a certain sense, if a musician were placed at each assigned point, the alpha point would be the point where one could best listen to the "harmony" produced by all the "musicians". Moreover, the alpha point detects the presumed outbreak source (impetus) of the distribution, the given input data set (the assigned points). Mathematically speaking, it is the opposite of the center of mass (centroid). In fact, given N points, the center of mass is the place from where all the points appear most similar to each other. Recall that the center of mass is based on the concept of average, and the average minimizes the sum of the square distances of all the points from the average itself. Consequently, the entropy of the distances in this place is maximal. The alpha point works in opposite way. The alpha point represents the place from where all the given points appear most different. For the same reason the alpha point is the place where the entropy is minimal and the free energy is maximal. Thus, the alpha point is the place from where all the given points have the maximal order and the maximal free energy.

A biological example can help. Let us imagine that the N given points represent the places from which a predator can come. Let us imagine also that a prey, knowing the positions of these points, needs to choose the safest place to have a rest. Considering these conditions, the alpha point represents, for the prey, the place from where it can control all the given points with the minimum of energy, because from this position they appear different from each other. Furthermore, the alpha point also represents, for the prey, the place where it maintains the maximum free energy it can spent. The inverse logic can be applied to this example. When we have N points where an unknown process is manifest, the alpha point can be said to be the most probable place from where the process itself originated (outbreak or impetus point). The more the alpha point is far away from the center of mass, the more the assigned points distribution is organized, and consequently the more "nonintuitive" the estimation of the outbreak or impetus point is when looking at the assigned point (input data) distribution. Figure 3.6 shows N assigned points ($n = 12$) and the position of alpha point and the center of mass.

3.4.2 *Alpha Map/Scalar Field*

The Alpha Map measures the degree of "activation" of different regions of the space in terms of their proximity, with respect to pseudo-distances, to the path leading to alpha star. The Alpha Map therefore estimates the short-term unfolding of the process by means of the observed distribution of past events. The associated equations are as follows (Fig. 3.7).

$$m_{kj} = \sqrt{\left(x_k - TWC_x(\alpha_j)\right)^2 + \left(y_k - TWC_y(\alpha_k)\right)^2}.$$

For Semantic TWC the equations changes as follows where we repeat the Euclidean distance with another variable name since it will be changed accordingly.

$$d_{kj} = \sqrt{\left(x_k - TWC_x(\alpha_j)\right)^2 + \left(y_k - TWC_y(\alpha_k)\right)^2}.$$

$$MeanD_i^{[CM]} = \frac{1}{N-1} \sum_{n}^{N} d_{kn}^{[CM]}$$

$$MeanD_i^{[Euclidean]} = \frac{1}{N-1} \sum_{n}^{N} d_{kn}^{[Euclidean]}$$

$$m_{lk} = \frac{\left(d_{kn}^{[Euclidean]}\right) \frac{1}{N} \sum_n^N MeanD_i^{[CM]}}{\frac{1}{N} \sum_n^N MeanD_i^{[Euclidean]}}$$

Fig. 3.7 Center of mass, path, distance

$$a_k = \frac{1}{N}\sum_{j=0}^{N-1} e^{-\frac{m_{kj}}{D}\beta^*},$$

where β^* (beta star) is defined below and the k is the index for each point of the plane and j is the index of each point of the TWC trajectory. The Alpha Map therefore estimates the short-term unfolding of the process by means of the observed distribution of past events.

Remark 3.25 The Alpha Map represents how much every grid point of the grid in which the process occurs is close to a "safe path", its alpha path. Because all the grid points cover the entire grid, the Alpha Map is a scalar field that may be represented with a "hot map", whose colors vary from dark red (grid points closer to the alpha path) to dark blue (grid points far away from the alpha path). If the resulting Alpha map is divided into intervals (bins) according to their color value, then it is possible to calculate the probability that each area of the map is selected in a random search. The less the probability is, the darker red area of the map, the more probable the outbreak point is location identified by the alpha point. The dark red area of the Alpha Map, in brief, indicates the average error of the position of the alpha point and it identifies the zone where the alpha point can be found. Because the alpha quantities detect the presumed origin or impetus of the assigned points distribution, then all these quantities linked to the alpha (alpha point, alpha path and Alpha Map) work as inferences connected to the past of the process represented by the assigned or input data points. Figure 3.8 shows the Alpha Map of the 12 point example, divided into 20 intervals (bins).

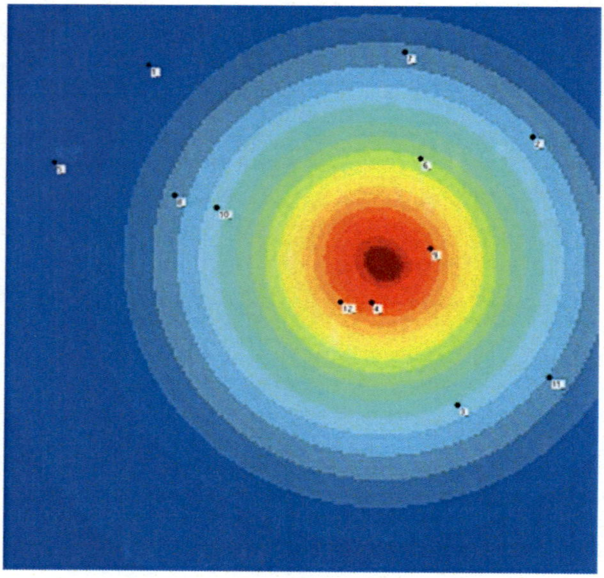

Fig. 3.8 Alpha map of the 12 point example

3.5 Self-Topological Weighted Centroid (STWC)

The self-topological weighted centroid points are pairs of coordinates (x, y) that form a trajectory similar to that of TWC-Alpha, but this time it considers the self attraction strength and defined below by Eqs. (3.35) and (3.36). The set of points (x, y) that, once drawn in a Cartesian plane, determine a trajectory whose starting point is the center of mass of the assigned points and the "optimal" point, the beta star (formally defined below) is determined by the maximum distance before the trajectory comes back towards the center of mass. The trajectory starts from the center of mass and stops when the maximum distance between the starting point and the point calculated by the iteration is attained. It is experimentally indicated that after a certain number of iterations, not known *a priori*, the trajectory curves and tends to return to the starting point since the strength of the point itself becomes dominant with respect to the others. The associated equations are as follows.

$$STWC_x(\beta_n) = \frac{1}{\sum_{i=1}^{N} w_i(\beta_n)} \sum_{i=1}^{N} w_i(\beta_n) x_i, \tag{3.35}$$

$$STWC_y(\beta_n) = \frac{1}{\sum_{i=1}^{N} w_i(\beta_n)} \sum_{i=1}^{N} w_i(\beta_n) y_i. \tag{3.36}$$

3.5.1 *Beta Star*

The beta star is the optimum control parameter of the STWC. It corresponds to the point where the trajectory stops. That is,

$$\beta^* = \max_{\beta_n} \left\{ dist\left(STWC_{\beta_n}, CenterMass \right) \right\} \tag{3.37}$$

The beta star parameter is of fundamental importance in the calculation of scalar fields (heat maps).

Remark 3.26 Given a distribution of N points in a plane, the optimal beta parameter (β^*) is the width of the bell curve preserving the interaction among the points. Under this optimal width each point tends to interact mainly with itself. This interaction is simulated considering, for each assigned point, a function of its self attraction from the other points, including itself. This quantity, weighted by the x and y of each point, is repeated in a loop, increasing the beta parameter monotonically. At each cycle a new pair of coordinates is generated (a new beta point). When the beta parameter is β_0 the beta point is the same as the center of mass. Increasing the beta parameter,

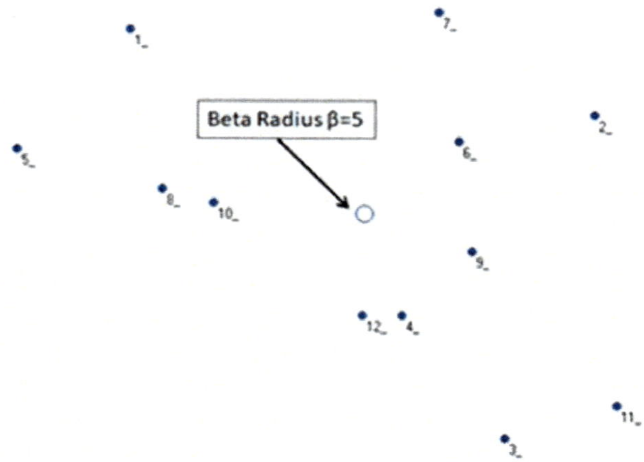

Fig. 3.9 Beta star for the 12 point example

the beta point moves away from the center of mass. But for a specific value of the beta parameter (depending on the distribution of the assigned points), the distance between the beta point and the center of mass stops increasing (in distance) and for bigger values of the beta parameter, the beta point subsequently starts to decrease in distance to the center of mass until the beta point is again the center of mass. When the increasing value of the beta parameter does not increase the distance between the beta point and the center of mass, it means that the interaction of each point with itself is becoming bigger than the interaction of that point with the others. Consequently, this value of the beta parameter, called Beta Star, is the limit over which the N assigned points stop to further influence the system of points. Figure 3.9 illustrates the optimal beta parameter.

3.6 TWC-Beta

The Beta Map defines the probability of observing new points (events) that belong to the same distribution of input data. It begins with the input data coordinates of the points in space from which the scalar field is calculated and the coordinates of the assigned/given data points. The associated calculations are as follows where the index i relates to the assigned or given data points and the index j relates to the grid points. The matrix of beta points is n_{ij} where

$$n_{ij} = \sqrt{\left(x_i - x_j\right)^2 + \left(y_i - y_j\right)^2}. \tag{3.38}$$

For the Semantic TWC this matrix is changed as follows. Here we repeat the distance between any grid point and the assigned point using another label since the beta points given by 3.38 will be modified.

$$d_{ij} = \sqrt{(x_i - x_j)^2 + (y_i - y_j)^2}$$

$$MeanD_i^{[CM]} = \frac{1}{N-1}\sum_n^N d_{kn}^{[CM]}$$

$$MeanD_i^{[Euclidean]} = \frac{1}{N-1}\sum_n^N d_{kn}^{[Euclidean]}$$

$$n_{lk} = \frac{\left(d_{kn}^{[Euclidean]}\right)\frac{1}{N}\sum_n^N MeanD_i^{[CM]}}{\frac{1}{N}\sum_n^N MeanD_i^{[Euclidean]}}$$

$$b_k = \frac{1}{N}\sum_{j=0}^{N-1} e^{-\frac{m_{kj}}{D}\beta^*},$$

and β^* is defined in Eq. (3.37) and the k is the index of each point of the plane and j is the index of each point of the TWC trajectory.

3.6.1 Beta Map

The Beta Map defines the probability of observing new points (events) that belong to the same distribution, that is, to the same data generating process. The Beta Map can therefore be regarded as a snapshot of a probability density function which is implicit in the structure of the data set. The output of the Beta Map is a $K \times K$ matrix, n_{ij}, $1 \le i, j \le K$. So in the end we have an activation value in the interval [0, 1] for each point of the grid.

Remark 3.27 The beta map represents a distribution function of other points, beyond the assigned points, at the time when the assigned points were collected. The beta map is a scalar field that may be visualized as a hot map: the darker red the area is, the higher is the probability to detect that there other points of the same process. In the epidemic analysis the red areas can point out new areas of interest where new cases can be found. The Beta map is regulated by three factors:
a. The beta star parameter to tune the influence that each one of the assigned points can have on the others;
b. The pseudo distance;
c. The attraction strength (w_i) that each point exercises in its neighboring area.
A biological example might help. Imagine that the N given points represent the places from which a predator can come. The beta map can be interpreted to represent the areas where the probability to find others predators is higher. In other words, the Beta map represents a map of probability to find new points at the present time. Figure 3.10 shows the beta map of the twelve point example.

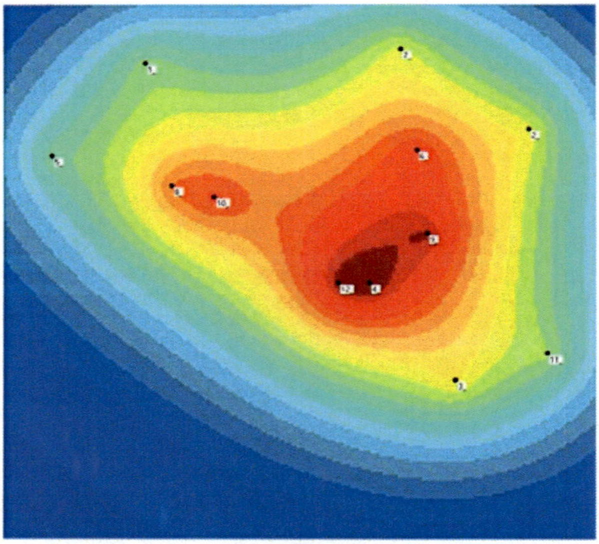

Fig. 3.10 Beta map for the 12 point example

3.7 TWC-Gamma

The gamma algorithm further probes into future configurations of the underlying structure of the process being studied.

3.7.1 *Gamma Paths*

Gamma paths are trajectories which, considering the attraction strengths exerted by each assigned point, connect the center of mass to each of them. Gamma paths consist of a finite number of points that varies from trajectory to trajectory. The control parameter is γ_i, for each of the source points. The input data consist of the coordinates of assigned points as well as the matrix of pseudo-distances. The output generates N different discrete trajectories, where N is the number of assigned points as before, connecting the center of mass to each point. The trajectory is calculated as follows.

$$p_{ij}(\gamma_i(t)) = e^{-\frac{d_{ij}}{D}\gamma_i(t)}.$$

The output is N different discrete trajectories, where N is the number of assigned points, connecting the center of mass to each point. The Semantic TWC gamma changes the above to

$$p_{ij}(\gamma_i(t)) = e^{-\frac{d_{ij}+d_{ij}^{[CM]}}{D}\gamma_i(t)}.$$

The trajectory are made up of the points

$$TWC_x(\gamma_i(t)) = \frac{1}{\sum_{j=1}^{N} p_{ij}(\gamma_i(t))} \sum_{j=1}^{N} p_{ij}(\gamma_i(t)) x_i,$$

$$TWC_y(\gamma_i(t)) = \frac{1}{\sum_{j=1}^{N} p_{ij}(\gamma_i(t))} \sum_{j=1}^{N} p_{ij}(\gamma_i(t)) y_i,$$

where D is the maximum of pseudo-distances and $\gamma_i(t+1) = \gamma_i(t) + \epsilon$. The main idea behind the creation of gamma trajectories is that at an early stage of the process, the points have somehow begun to "communicate" with each other. Since the system is not yet organized or optimized in principle, the "messages" sent take on random trajectories, thus favoring the passage through the center of mass. For this reason, it has been hypothesized that "communication" in the first future will take place using the center of mass as an intermediary.

Remark 3.28 The gamma scalar field may be thought of as a further qualification of the beta field, which transforms attraction strengths into intensities of network interaction. In fact, it may be regarded as a "prediction" of the evolution of the beta field, when the points start communicating in a non-organized way. The gamma map needs that a "simulation of messages" exchange among the assigned points is introduced. For this reason gamma map may be considered the immediate future of the beta map. Consequently, gamma map is hypothesized to have a predictive nature. In a situation where no point "knows" the position of the other, in order to optimize the path of its messages to the first neighbors, the center of mass tends to become the first addressee of all the points. Consequently, it is useful to assume that in this phase the center of mass works as the central hub where all the messages are distributed from each point to any other point. The path linking each point to the center of mass can be simulated taking account all the other paths, in order to generate a systemic picture where each path is influenced by all the others. Figure 3.11 shows the gamma paths between the twelve point example and the center of mass.

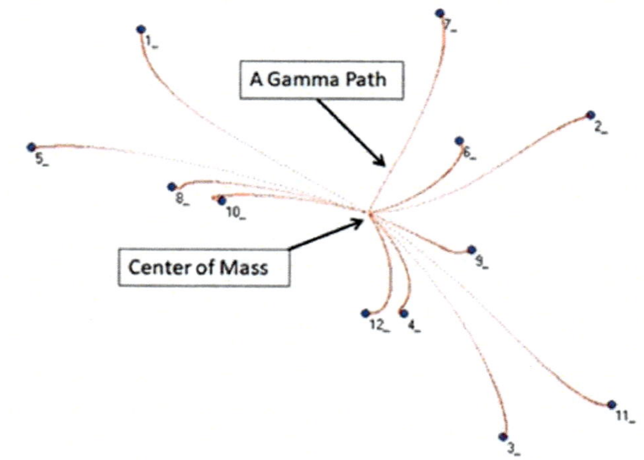

Fig. 3.11 Gamma path of the 12 point example

3.7.2 *Gamma Map*

The gamma map is the scalar field generated calculating the distance of each grid point of the box from all the gamma paths. The gamma map, or gamma scalar field, describes the extent to which each point of the space is activated by its closeness to any of the points belonging to any of the $TWC(\gamma_i(t))$ trajectories. The input to this part of the algorithm consists of all the gamma paths, that is, the N vector of coordinates

$$TWC(\gamma_i(t)) = \left\{\left(TWC_x(\gamma_i(t)), TWC_y(\gamma_i(t))\right)\right\}, t = 1, 2, ...Q_i.$$

A space U, usually a square, is a subset of the plane on which to the scalar field is defined. The trajectory is calculated as follows.

$$m_{k\gamma_i(t)} = \sqrt{(x_k - TWC_x(\gamma_i(t)))^2 + \left(y_k - TWC_y(\gamma_i(t))\right)^2},$$

For Semantic TWC the change is

$$d_{k\gamma_i(t)} = \sqrt{(x_k - TWC_x(\gamma_i(t)))^2 + \left(y_k - TWC_y(\gamma_i(t))\right)^2}$$

$$MeanD_i^{[CM]} = \frac{1}{N-1}\sum_n^N d_{kn}^{[CM]}$$

$$MeanD_i^{[Euclidean]} = \frac{1}{N-1}\sum_n^N d_{kn}^{[Euclidean]}$$

$$m_{k\gamma_i(t)} = \frac{\left(d_{k\gamma_i(t)}\right)\frac{1}{N}\sum_n^N MeanD_i^{[CM]}}{\frac{1}{N}\sum_n^N MeanD_i^{[Euclidean]}}$$

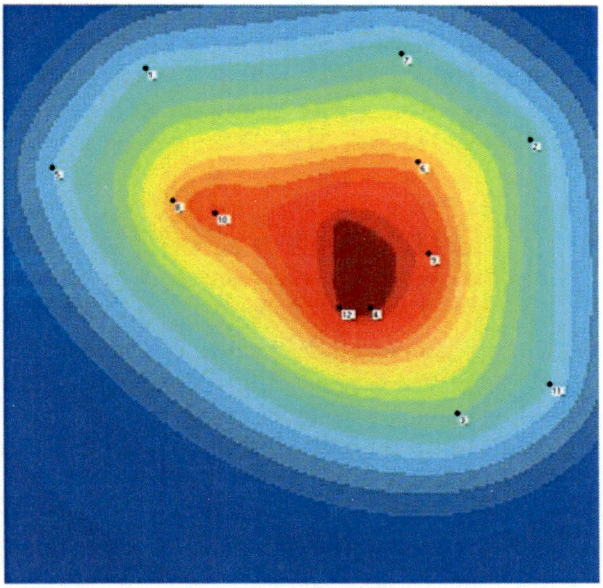

Fig. 3.12 Gamma map for the 12 point example

$$c_k = \frac{1}{\sum_{i=1}^{N} Q_i} \sum_{i=1}^{N} \sum_{j=0}^{V-1} e^{-\frac{m_{k\gamma_i}(t)}{D}\beta^*},$$

where the points (x_k, y_k) are points of U and Q_i is the number of points making up the ith trajectory. The gamma map for the 12 point example is given next in Fig. 3.12.

3.8 TWC-Theta

TWC—Theta consists of the theta path, theta distances, and uses the minimal spanning tree to depict the connections between the assigned points. These connections can be interpreted as the possible pathways of the underlying process being modeled. We begin with the theta path.

3.8.1 Theta Paths

The $N(N-1)/2$ gamma paths that connect directly each pair of points of the data set, according to their nonlinear distances from the center of mass consists of what we call the *theta path*. In this way, for each possible interaction between any couple of points, we can determine, on the basis of the curvature of the corresponding nonlinear edge, what is the spatial direction that exerts the most significant interference on that interaction. What is needed is the gamma paths as input and the coordinates of assigned points. What results for the algorithm are the $N(N-1)/2$ trajectories connecting each assigned point with all the others. It is a complete regular graph with nonlinear edges. These are calculated as follows.

$$TWC\_\theta_{(ij)_x}(t) = x_j + \left(TWC_{i_x}(\gamma_i(t)) - TWC_{j_x}(\gamma_i(t))\right), \tag{3.39}$$

$$TWC\_\theta_{(ij)_y}(t) = y_j + \left(TWC_{i_y}(\gamma_i(t)) - TWC_{j_y}(\gamma_i(t))\right). \tag{3.40}$$

After an initial random interaction, it is assumed that the points start to "communicate" (be related or connect) in a more direct and organized way. For this reason, based on gamma trajectories, the process is thought to evolves towards a further future that directly connects each point with each other without the need to pass through the center of mass.

3.8.2 Theta Distances

A matrix that contains all the nonlinear distances between all the assigned points is needed as input. It quantifies all the nonlinear connections (interactions) between the points of the spatial distribution. What we call the *theta distances* consists of a matrix that contains all the nonlinear distances between all the assigned points. It quantifies all the nonlinear connections, that is, "interactions" between the points of the spatial distribution. What is needed for the algorithm are the theta paths. The results is the $N \times N$ matrix, where N is the number of the assigned points. The algorithm calculates this in the following way.

$$\theta_{ij} = \sum_{t=1}^{Z_{ij}-1} dist\left(TWC\_\theta_{(ij)_x}(t), TWC\_\theta_{(ij)_y}(t)\right),$$

$$Z_{ij} = \min\left\{Q_i, Q_j\right\},$$

where Q_i, Q_j are the number of points of gamma paths and $dist(\_, \_)$ is the usual Euclidean distance between two consecutive points of the same theta path, those defined by Eqs. (3.39) and (3.40). The theta distance arises from the need to measure distances between points not only in Euclidean terms but taking into account past history and the way they interact. It is therefore based on the length of the theta trajectories that join a pair of points knowing that these take into account the gamma paths.

3.8.3 Nonlinear Minimum Spanning Tree

The nonlinear minimum spanning tree (NL-MST) represents the minimum-energy tree configuration that connects all the points of the distribution, pruning away all the edges that are not indispensable to maintain the connectedness among points by means of the most relevant interactions. The tree requires theta distances and Euclidean distances and produces the NL-MST with nonlinear edges. The calculation algorithm is provided with the following characteristics.

1. The MST is calculated from the Euclidean distance matrix using any well-known Kruskal algorithm (see [3]).
2. The list of linear edges chosen by the algorithm is selected.
3. The linear edges are replaced by nonlinear theta trajectories.

Thus, TWC uses two associated algorithm that are typically used in a graph theoretic and in the artificial adaptive systems setting (see [2]). For the context of TWC, the MST is used in conjunction with TWC-Theta, in which the associated graph between the nodes (geographic locations of the given data) indicating paths between the nodes of greatest significance is generated. Recall that MST finds an acyclic subset T of nodes of all edges E between the geographical locations given by the data that connects all of the vertices in the graph whose total weight is minimized, where total weight in our case is given by

$$D(T_r) = \sum_{k=0}^{N_v-1} \hat{d}_{i_{r_k} j_{r_k}}, \tag{3.41}$$

and N_v is the number of vertices (geographic locations given by the data), T_r is called the rth tree. The MST is the T_{r^*} tree whose weighted sum of edges attains the minimum value, that is,

$$MST = T_{r^*} = \min \{d(T_r)\}. \tag{3.42}$$

Kruskal [3] found an algorithm to determine the MST of any undirected graph in a quadratic number of steps for the worst case. Obviously, the Kruskal algorithm generates one of the possible MSTs but not all of them. The connections for the 12-Point example is depicted in Fig. 3.13.

The NL-MST for the 12-Point example is depicted in Fig. 3.14.

3.8.4 Theta Map

The theta map that is calculated contains the future of the gamma map. What is required for the algorithm is the NL-MST, and a space U, usually a square and

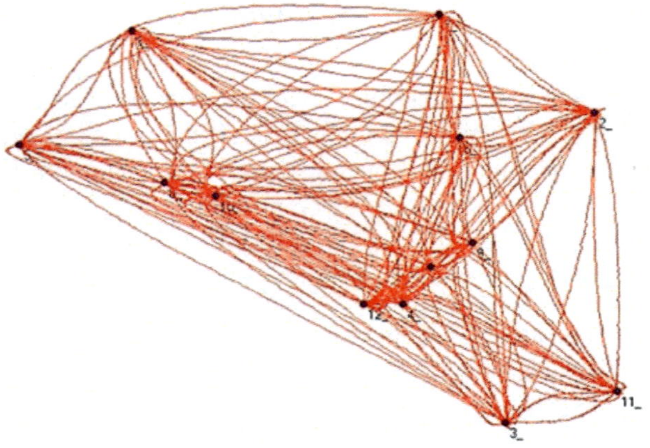

Fig. 3.13 Nonlinear connection for the 12-point example

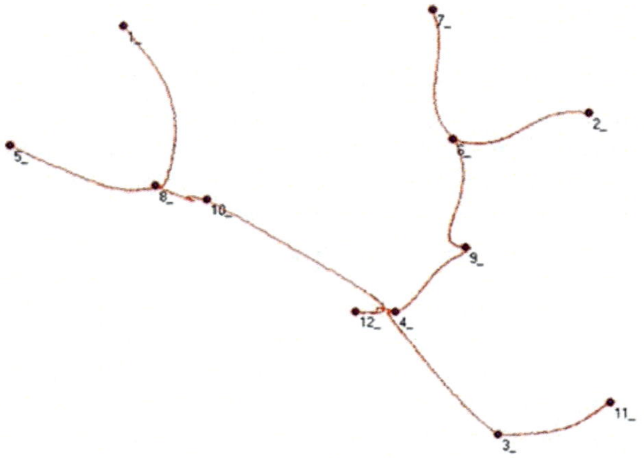

Fig. 3.14 NL-MST for the 12-point example

subset of the plane, on which to define the scalar field. Also, the beta star, and D, the maximum of pseudo-distances, are also needed. The calculations are as follows.

$$I_{k\theta_z(t)} = \sqrt{\left(x_k - DirectPath\_\theta_{z_x}(t)\right)^2 + \left(y_k - DirectPath\_\theta_{z_x}(t)\right)^2}$$

and for Semantic TWC we have

$$d_{k\theta_z(t)} = \sqrt{\left(x_k - DirectPath\_\theta_{z_x}(t)\right)^2 + \left(y_k - DirectPath\_\theta_{z_x}(t)\right)^2}.$$

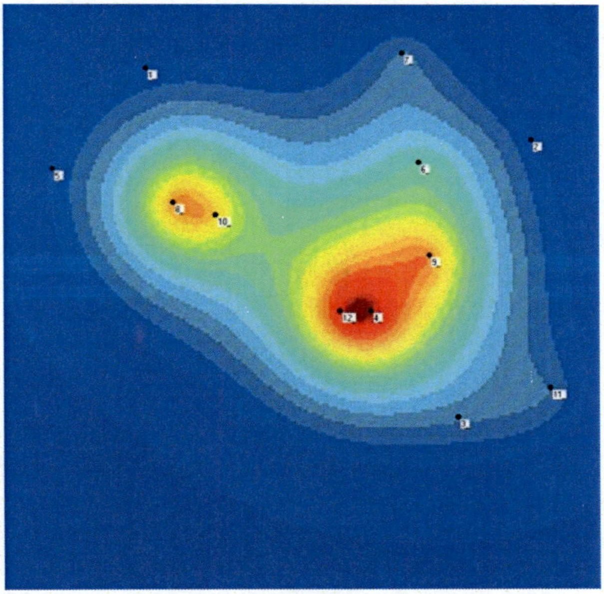

Fig. 3.15 Theta map for the 12-point example

$$MeanD_i^{[CM]} = \frac{1}{N-1}\sum_{n}^{N} d_{kn}^{[CM]}$$

$$MeanD_i^{[Euclidean]} = \frac{1}{N-1}\sum_{n}^{N} d_{kn}^{[Euclidean]}$$

$$I_{k\theta_z(t)} = \frac{\left(d_{k\theta_z(t)}\right)\frac{1}{N}\sum_{n}^{N} MeanD_i^{[CM]}}{\frac{1}{N}\sum_{n}^{N} MeanD_i^{[Euclidean]}}$$

$$e_k = \frac{1}{\sum_{z=1}^{N} P_z}\sum_{z=1}^{N-1}\sum_{j=0}^{P_z} e^{-\frac{m_{k\gamma_i(t)}}{D}\beta^*},$$

where the points (x_k, y_k) belong to U and P_z is the number of points composing the edge z of the NL-MST from the $DirectPath\_\theta_z$. The Theta Map for the 12-Point example is depict in Fig. 3.15.

3.8.5 Theta Transition Matrix via Markov Chains

The theta transition matrix via a Markov chain (see [4]) is a probability matrix based upon the theta distance matrix. The calculation of the transition matrix is obtained as follows.

$$\left\{\tilde{\theta}_{ij}\right\}_{ij} = \begin{cases} \tilde{\theta}_{ij} = \max\_\theta - \theta_{ij} & \text{if } i \neq j \\ \tilde{\theta}_{ij} = 1 & \text{if } i = j \end{cases},$$

$$P_{ij} = \frac{\tilde{\theta}_{ij}}{\sum_{k=1}^{N} \tilde{\theta}_{ij}}$$

where $\max\_\theta - \theta_{ij}$ is the maximum theta distance among the points. The transition matrix estimates the probability that any given point in the spatial distribution "communicates" (relates, connects, interacts) with another one as the effect of some random external shock (for instance, a local shortage of some key resource for quasi-epidemic processes, or the occurrence of a mutation of the infective agent in an epidemic process). The closer two points in the spatial distribution (according to theta), the more likely that the random shock will force an interaction between them. This assumption can be taken as the basis for building an explicit diffusion model for our spatial distribution.

3.8.6 Discrete Time Markov Chain (DTMC)

The DTMC is an explicit dynamical model that allows the computation of the attractor of the dynamics. It uses the theta transition matrix as input and produces a graph of attractors. The components for the calculation of the DTMC algorithm are as follows.

1. The number of calls to be made is fixed.
2. One of the N assigned points is randomly extracted.
3. Another point and a pseudo-random number are randomly extracted.
   (a) If the pseudo-random number is greater than the transition probability the algorithm returns to point 2.
   (b) If the pseudo-random number is less than the probability of transition, the passage takes place and the algorithm proceeds from point 3 taking as its first point the last extract.
4. A statistic of the most frequent transitions and attractor points, that is, from which one never manages to exit, is carried out.

3.9 TWC—Iota

The final algorithm of the TWC suite of algorithms to be discussed is TWC - Iota which is presented next.

3.9.1 Iota Projected Points

Each originally assigned point is projected onto the map. Each new point denotes the possible position of a new point in the future. To compute the iota projected points, the selected matrix of meta-distances, according to one of the previous criteria, and the a multi-dimensional scaling (MDS) algorithm, are needed. The results are outputted as a set of new N points where (x, y) are the coordinates of each one. The selected matrix is projected into a $2 - dimensional$ space by means of the MDS algorithms. If the matrix is $N \times N$, then N new points corresponding to those (meta) distances will be obtained. Clearly the positions of the projected points depend on the selected matrix. What is outputted are the (x, y) coordinates of meta-clusters.

3.9.2 Meta-Distance in TWC—Iota

The general topic of meta-distance has been previously mentioned. Recall that a meta-distance matrix associated with N points is the distance matrix of their distances. The input is an $N \times N$ distance matrix where $D = \left(d_{ij}\right) = \left(d_{ij}^{[0]}\right)$ is the Euclidean distance from i to j. Then,

$$d_{ij}^{[0]} = \sqrt{\sum_{k=1}^{N} \left(d_{ik} - d_{jk}\right)^2}.$$

The idea is quite simple. It aims to measure how far the distances between each point and all the others are.

Iterative Meta-Distance Matrices

The iterative meta-distance matrix is

$$d_{ij}^{[t+1]} = \sqrt{\sum_{k=1}^{N} \left(d_{ik}^{[t]} - d_{jk}^{[t]}\right)^2}, t = 0, 1, ..., n .$$

This calculation of meta-distance will diverge. Thus, there is a convergent iterative meta-distance that is given next. We start with the iterative meta-distance

$$D_{ij}^{[t+1]} = \sqrt{\sum_{k=1}^{N} \left(d_{ik}^{[t]} - d_{jk}^{[t]}\right)^2}, t = 0, 1, ..., n \ . \tag{3.43}$$

$$d_{ij}^{[t+1]} = f(D_{ij}^{[t+1]}). \tag{3.44}$$

For the Sematic TWC Eq. (3.44) is modified to be

$$d_{ij}^{[t+1]} = \frac{1}{2} f(D_{ij}^{[t+1]}) + \left(1 - w_{ij}^{[CM]}\right), 0 < w_{ij}^{[CM]} \leq 1$$

$D_{ij}^{[0]}$ is the usual Euclidean distance matrix,

$$f(D_{ij}^{[t]}) = scale\left(D_{ij}^{[t]}\right) + offset.$$

Note that $f(D_{ij}^{[t]})$ is a function that linearly scales the input in the range of [0, 1] for

$$scale\left(D_{ij}^{[t]}\right) = \frac{1}{\max D_{ij}^{[t]} - \min D_{ij}^{[t]}},$$

$$offset = \frac{\min D_{ij}^{[t]}}{\max D_{ij}^{[t]} - \min D_{ij}^{[t]}}.$$

The sequence will converge since we are iterating on numbers between 0 and 1 (at zero and one it stays at zero and one). Thus, one can stop the iteration when

$$\sum_{i,j} \left| d_{ij}^{[t+1]} - d_{ij}^{[t]} \right| < \ \in$$

where $\in$ is small.

Remark 3.29 Given a distance matrix, it is always possible to consider the rows (or columns) as vectors and calculate again a distance matrix of the same size. In the case that the distance matrices are iteratively scaled in the range [0,1] convergence is ensured. Moreover, it is experimentally verified that clusters are created on the matrix while letting the system evolve until convergence.

Meta-Distance Matrix Selection

The iterative converging method for meta-distances produces a finite family of matrices. Thus, it is necessary to establish a criterion for selecting one. This is discussed next. The input is a family of meta-distance matrices. We will output a single meta-distance matrix. A first meta-distance matrix is selected and it is denoted $D_{ij}^{[1]}$ and a last meta-distance matrix is selected which is denoted $D_{ij}^{[K]}$. What we call the delta delta J matrix criterion is given as follows.

1. *For* $t = 2, ..., K$ *do steps* $2 - 7$.
2. $S^{[t]} = \sum_{i=1}^{N-1} \sum_{j=i+1}^{N} D_{ij}^{[t]}$,
3. $\Delta S^{[t]} = \sum_{i=1}^{N-1} \sum_{j=i+1}^{N} \left(D_{ij}^{[t]} - D_{ij}^{[t-1]} \right)$,
4. $\begin{Bmatrix} \delta = 1 & \Delta S^{[t]} > 0 \\ \delta = -1 & \Delta S^{[t]} < 0 \end{Bmatrix}$,
5. $J^{[t]} = \log \left(\frac{S^{[t]}}{|\Delta S^{[t]}|} \right) \delta$,
6. $\Delta J^{[t,t-1]} = J^{[t]} - J^{[t-1]}$,
7. $\Delta\Delta J^{[t,t-1,t-2]} = J^{[t,t-1]} - J^{[t-1,t-2]}$,
8. $t^* = \max_t \left\{ \Delta\Delta J^{[t,t-1,t-2]} \right\}$,
9. $M_{ij} = D^{[t^*]}$.

3.9.3 Meta-Clusters

Once the new points have been projected, one can determine the main clusters of the distribution. These points correspond to the strategic areas for the diffusion of the process that is being analyzed. The algorithm needs the coordinates of new projected points and a clustering algorithm. The results output the (x, y) coordinates of meta-clusters. The clustering algorithm is applied to the new points. In order to get the centroids, the following are needed.

- The $k - Means$ algorithm is used for each k from 2 to N, where N is the number of points.
- For each iteration of the $k - Means$ algorithm, the Davies-Boulding index, [5], is computed.
- The iteration relevant to the best index is chosen, so that an optimal k is defined, k^*.
- The centroid of each of the k^* clusters is considered as one of the meta-cluster.

Remark 3.30 It is possible to interpret the points corresponding to the meta-clusters as strategic points for the diffusion of the studied process.

The iota points represent where the N assigned points tend to go, when, according to their distribution, they move in order to cluster. In 2−dimensions the iota points tend to converge to their vanishing points. Thus, the vanishing points in the context of epidemics might indicate where the last cases of the epidemic will be located.

This means that the iota points tend to converge to their vanishing points. In 2−dimensions there will be two convergent vanishing points, and step by step all the assigned points will collapse in two clusters. Consequently, the migration of the assigned points toward their vanishing points describes a vectorial field. At the end

Fig. 3.16 Locations of the 12-point example

Fig. 3.17 Migration of the !2-point example

of this process, each one of the assigned points will show many replicas of itself, each one in a different position from its original position. The entire population of these points we call *iota points*.

It is important to note that the iota points can be placed outside of the convex hull of the assigned points. This feature is fundamental in order to understand the capability of the iota points to also detect new areas of interest outside the original convex hull area of the assigned points. Figure 3.16 shows the N ($n = 12$) assigned points for he 12-Point example and Fig. 3.18 shows the iota points generated from the first ones, during their migration (Fig. 3.17).

Remark 3.31 The iota algorithm is based on the iterative calculation of the distances among the assigned points. At each iteration a new meta-distance (distance of a distance) is computed. Because the distance in two dimensions is calculated on the hypotenuse of a right triangle, then at each iteration the distance should increase. However, if after each iteration, the new distance metric is linearly scaled between 0 and 1, then after a finite number of iteration the iota algorithm converges to a fixed

Fig. 3.18 Clusters for the 12-point example

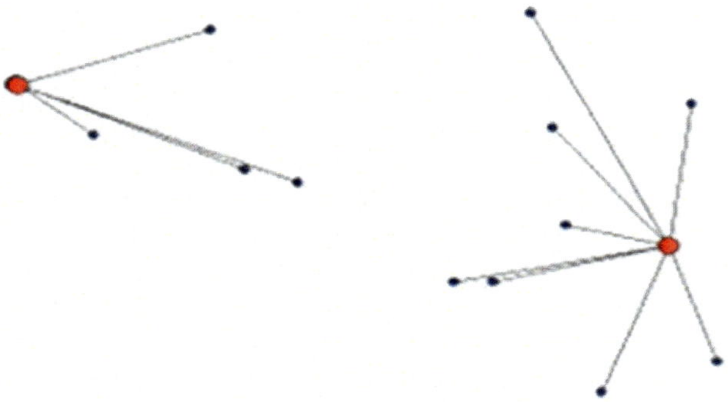

Fig. 3.19 Final two clusters for the 12-point example

point, generating two big clusters where each of the assigned points will collapse. Figure 3.19 shows the final step of this process, when all the assigned points collapse into two clusters.

The iota algorithm ends its loop of iterations when the matrix of distances at iteration (n) is equivalent to the matrix of distances at iteration ($n - 1$); that is, it does not change anymore. The iota algorithm does not consider as significant the position of replicas of the assigned points, only when the series converges. The iota algorithm considers a distribution of replicas of the assigned points a salient distribution when the summation of delta delta of the distances at iteration (n) and iteration($n - 1$) is the biggest. This happens when the biggest change of regime occurs between a matrix of distances and the previous one.

Figure 3.20 shows the dynamics of the deltas (the values of the increment of changes) after any iteration. For our 12-Point Example, this occurs at iteration #14. There is a peak that represents the most salient distribution of replicas of the assigned points. Figure 3.21 shows how each assigned point is projected into its replica at iteration #14. Figure 3.22 shows, finally, the two centroids of the clusters generated by the all the replicas of the assigned points.

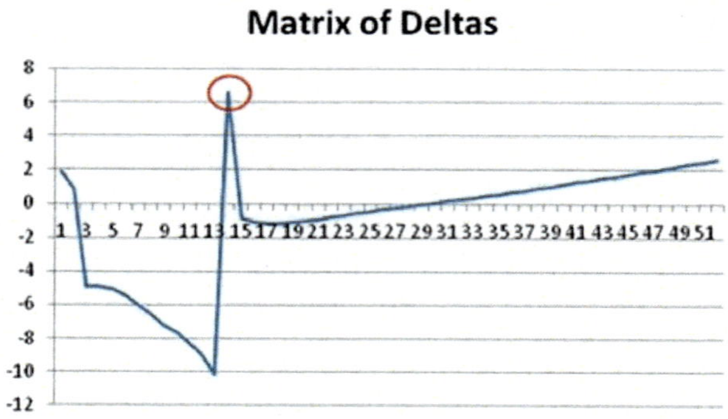

Fig. 3.20 Dynamics of the delta delta matrices for the 12-point example

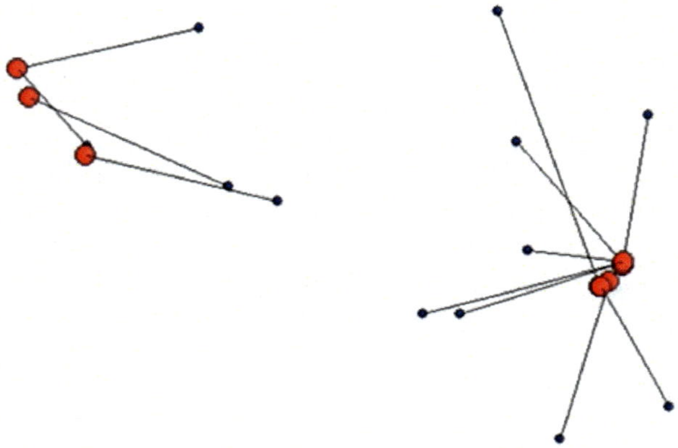

Fig. 3.21 Salient Iota replica (Red) of each assigned point (Blue) at iteration 14

Fig. 3.22 Assigned points (Blue), salient clusters (Red), iteration 14

Cluster #1	
Name	Fuzzy Clustering Membership Value
ID_2	0.7948
ID_3	0.7771
ID_4	0.7513
ID_6	0.7522
ID_7	0.6248
ID_9	0.8576
ID_11	0.8249
ID_12	0.6965
Cluster #2	
ID_1	0.7275
ID_5	0.8798
ID_8	0.6456
ID_10	0.5616

Membership Values of the Two Clusters

3.9.4 Fuzzy Membership to the Meta-Clusters

Through the clustering algorithm each point is assigned to one of the meta-clusters. Besides the meta-clusters, the coordinates of meta-clusters are needed and for each of the coordinates, a (fuzzy) membership is calculated. This (membership) value for each of the points and each of the meta-clusters is computed. The calculations is as follows.

$$\mu_{c_k}(p_i) = 1 - \frac{d_{ic_k}}{\sum_j^G d_{ic_j}}$$

where

$$\begin{aligned}
\mu_{c_k}(p_i) &= \text{the membership of the } i\text{thpoint with respect to the } k\text{th cluster,} \\
d_{ic_k} &= \text{distance between the } i\text{thpoint and the centroid of the } k\text{th cluster,} \\
d_{ic_j} &= \text{distance between the } i\text{thpoint and the centroid of the } j\text{th cluster,} \\
G &= \text{the number of clusters.}
\end{aligned}$$

Remark 3.32 Through the fuzzy membership function it is possible to quantitatively evaluate how much each point belongs to one of the two clusters. This procedure is general, but it's usually computed to know how much each projected point is typical of its centroid.

Iota points and iota clusters can be considered to represents the virtual dynamics that the assigned points could manifest when a perturbation is activated at the center of mass of the points distribution. A biological example of this dynamics could be represented by a herd of herbivores (assigned points) whose center is suddenly perturbed and consequently the herd is pushed to split in two clusters. The centroids of this new clusters (iota clusters) are the best positions for predators to attack the herd after the perturbation.

3.9.5 Iota Map

The iota map is generated by the distance of each generic point of the two-dimensional space from each projected point of the meta-distance algorithm, weighted by the distance that each projected point has from its original point. The farther away a projected point is from its original point, the stronger the weight. The coordinates of assigned points (P_S), the coordinates of projected points (P_P), the coordinates of the points in space U, from which one wants to calculate the scalar field (P_g), and the maximum of pseudo-distances are needed as inputs for this algorithm. What this produces is a scalar field that is represented by a heat map. The calculations are done as follows.

$$Md_k = \sqrt{\left(P_{S_{k(x)}} - P_{P_{k(x)}}\right)^2 + \left(P_{S_{k(y)}} - P_{P_{k(y)}}\right)^2},$$

$$Dw_i = \sum_{k=1}^{N} \left(1 - \frac{Md_k}{\sum_{j=1}^{N} Md_k}\right) \frac{\sqrt{\left(P_{P_{k(x)}} - P_{g_{k(x)}}\right)^2 + \left(P_{P_{k(y)}} - P_{g_{k(y)}}\right)^2}}{MaxD},$$

$$A_i = e^{-Dw_i},$$

where $MaxD$ is the maximum distance among the assigned points.

Remark 3.33 TWC—Iota takes into account how each point is projected on the map. For this reason, in the case of processes related to epidemics, it is interpreted as the level of infectivity of the different areas.

The iota map represents the main area of clustering of the assigned points. In order to understand the meaning of the iota map, it is important to consider two distances:

- The distance between each assigned point and its projection into an iota point;
- The distance between each iota projection point and any generic grid point.

Figure 3.23 shows these key distances.

The iota map is the results of the distance of each grid point from all iota points, where each iota point is weighted by its distance from the assigned point to those

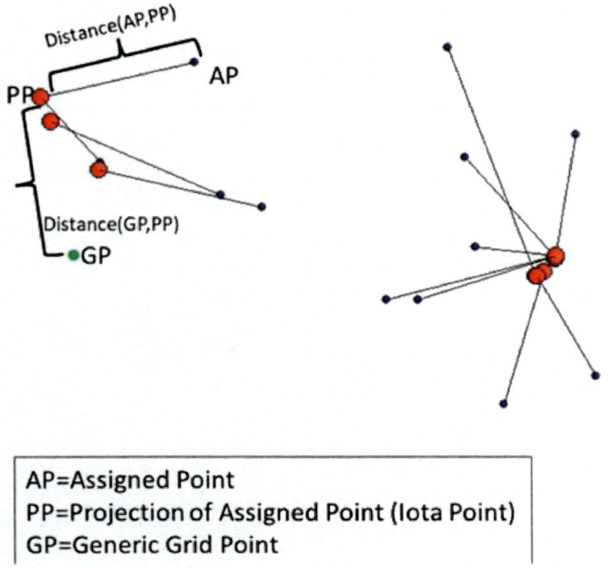

Fig. 3.23 Iota map of two distance measures

of its projection. The iota map appears as a "hot map" in which its dark red area is far away for all the assigned points. The iota map points out a new zone from where or toward where the assigned points tend to come or tend to go. The iota algorithm, in practice, could detect a critical piece of information about the past or about the future of the system of points. The forward or the backward direction of its meaning depends on the previous history of the assigned points. Figure 3.24 shows the iota map.

3.9.6 Meta-Beta

Once the plane has been repopulated with new possible future points, it is possible to reapply the algorithm to obtain the TWC-Beta in order to have the estimate of the current distribution, in a hypothetical future. The inputs of this algorithm are the coordinates of the points in the space U from which one wants to calculate the scalar field, the coordinates of assigned points, the coordinates of the projected points (hereafter both are considered as new assigned points), the maximum Euclidean distance among the new assigned points, the vector of non-parametric attraction strength computed considering the new assigned points, and the beta star relevant to the new assigned points. The output is a $K \times K$ matrix where $K \times K$ is the number of points of the space U. Therefore, in the end one will get an activation value in [0, 1] for each point in the considered space. The calculations are as follows.

Fig. 3.24 Iota map for the 12-point example

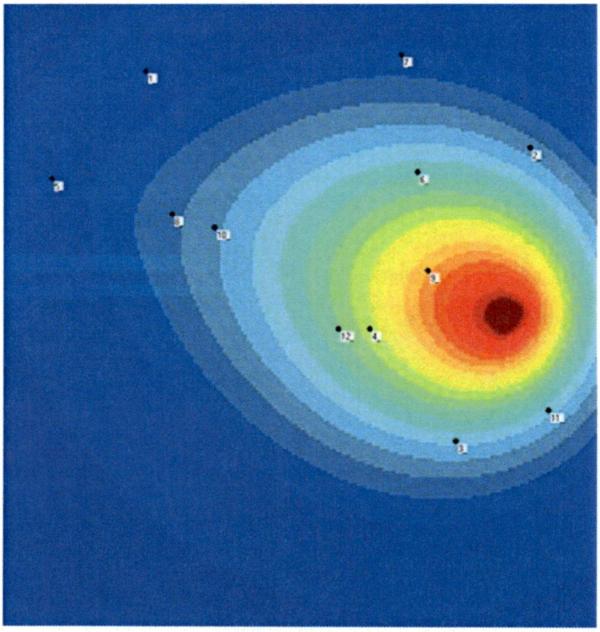

$$n_{kj} = \sqrt{\left(x_k - x_j\right)^2 + \left(y_k - y_j\right)^2},$$

$$b_k = \frac{1}{N}\sum_{j=1}^{N} w_j e^{-\frac{n_{jk}}{D}\beta^*}$$

where k is the index for each point of the plane and j is the index for each new assigned point.

Remark 3.34 Thus, meta-beta allows a projection of the possible future evolution of the process. Clearly, in addition to the TWC Meta-Beta, it is possible to replicate all the other types of the TWC theory (Meta-Gamma, Meta-Theta, Meta-Iota).

3.10 Summary

A brief written and visual summary of the concepts developed in this is useful at this point. TWC is an adaptive algorithm that generates a set of quantities able to transform a static distribution of points in the planar space into a system of points with a past and future history. The following is a brief list of these quantities.

1. Pseudo Distance (metric of the system of points)
2. Alpha Point, Alpha Path and Alpha Map (the past of the system of points)

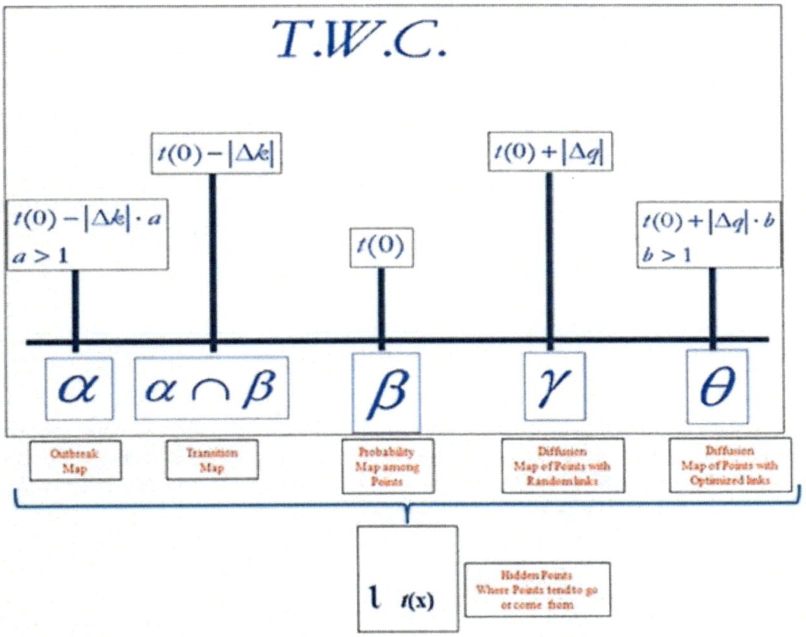

Fig. 3.25 A synthesis of TWC items

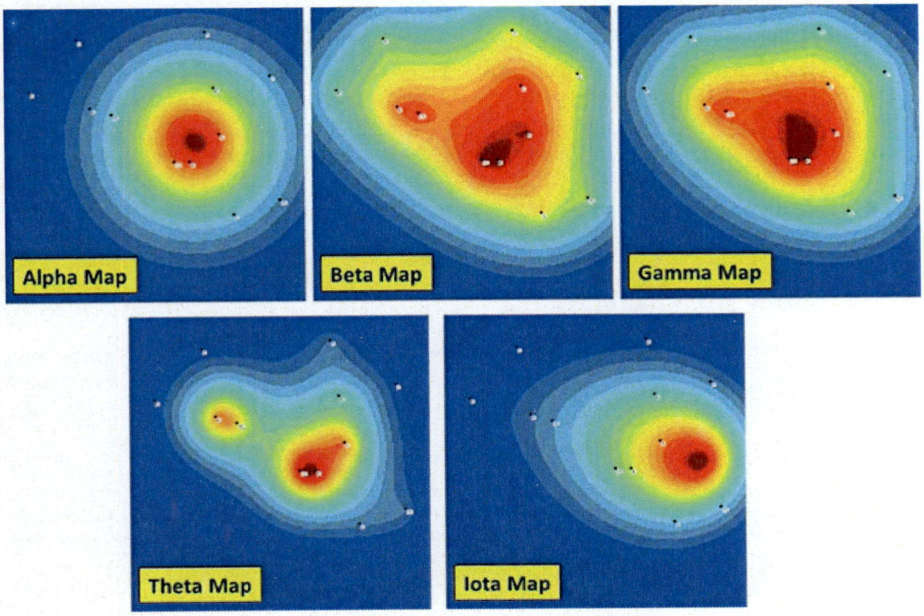

Fig. 3.26 TWC maps

3. Beta Parameter and Beta Map (the present of the system of points)
4. Gamma Paths and Gamma Map (the immediate future of the system of points)
5. Theta Nonlinear MST and Theta Map (the medium future of the system of points)
6. Iota Points, Iota Clusters and Iota Map (the final future or first origin of the system of points)

Figure 3.25 below synthesizes these items. Figure 3.26 shows the different maps produced by TWC in the points distribution analyzed for the 12-Point Example.

References

1. M. Buscema, M. Breda, E. Grossi, L. Catzola, P.L. Sacco, Semantics of point spaces through the topological weighted centroid and other mathematical quantities: theory and applications, Chapter 4, in *Data Mining Applications Using Artificial Adaptive Systems*, ed. by W.J. Tastle (Springer Science+Business Media New York, 2013), pp. 75–139
2. M. Buscema, W. Lodwick, M. Breda, G. Massini, F. Newman, M.A. Zeydabadi, in *Artificial Adaptive Systems Using Auto Contractive Maps: Theory*. Applications and Extensions (Springer, 2018). https://doi.org/10.1007/978-3-319-75049-1
3. J. B. Kruskal, On the shortest spanning subtree of a graph and the traveling salesman problem, *Proceedings of the American Mathematical Society*, vol. 7, Num. 1, pp. 48–50 (1956)
4. B. Bolch, S. Greiner, H. de Meer, K.S. Trivedi, K.S., **Queueing Networks and Markov Chains**, 2nd ed. (Wiley, New York, 2006)
5. D.L. Davies, W. Boulding, A clustering separation measure. IEEE Trans. Pattern Anal. Mach. Intell. **1**, Num. **2**, 224–227 (1979)

Chapter 4
Illustrative Examples

Two simple examples are presented next to further illustrate the TWC algorithms so that a sense of how the algorithms analyze geographic data work. The example of Chap. 3 came from an application. The following two examples are made up. The first example consists of 10 points and is one in which there are two points that are "outliers". The second example consists of 11 points and has a more random distribution set of points. We use the equations developed in Chap. 3 and apply them to these two examples using MATLAB whose code can be found in the Appendix.

4.1 Two Simple Example Data Sets

The two example data sets are presented next followed by their corresponding TWC maps.

Example 4.1 Example data set 1 of 10 points (with two "outliers")

$$\begin{array}{lrr} Data & x & y \\ P_1 & 2 & 11 \\ P_2 & 2 & 13 \\ P_3 & 35 & 14 \\ P_4 & 28 & 14 \\ P_5 & 30 & 15 \\ P_6 & 34 & 10 \\ P_7 & 35 & 12 \\ P_8 & 33 & 16 \\ P_9 & 31 & 10 \\ P_{10} & 28 & 12 \end{array} \tag{4.1}$$

is depicted in Fig. 4.1, which shows the coordinates of points.

M. Buscema et al., *The Topological Weighted Centroid: A New Vision of Geographic Profiling*, Studies in Computational Intelligence 1095,
https://doi.org/10.1007/978-3-031-28901-9_4

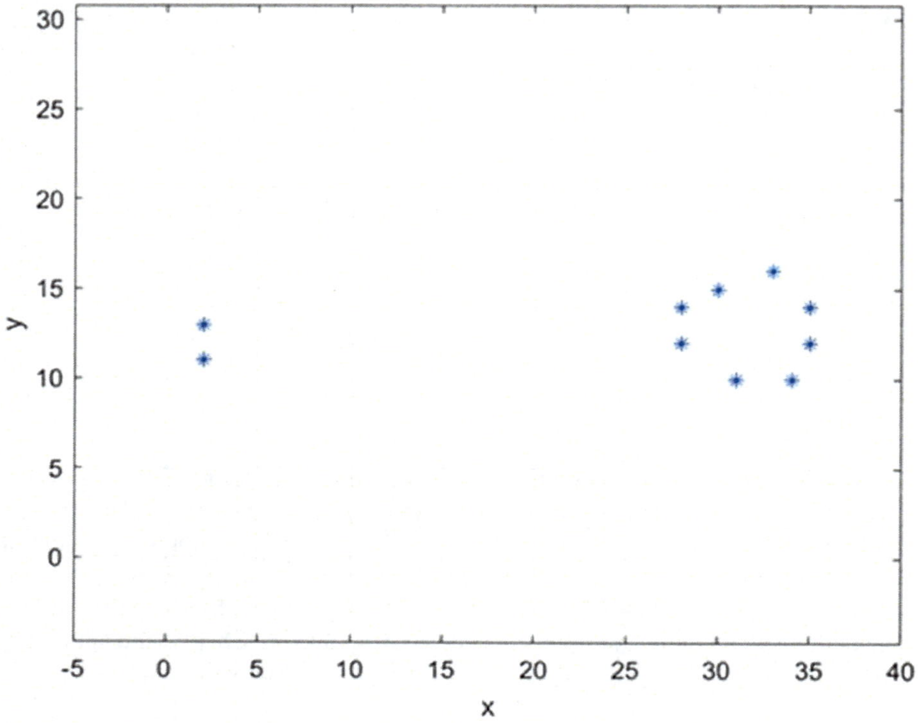

Fig. 4.1 10 point illustrative Example 4.1

Example 4.2 Example data set 2 of 11 points

$$\begin{array}{ccc} Data & x & y \\ P_1 & 3 & 14 \\ P_2 & 13 & 25 \\ P_3 & 2 & 8 \\ P_4 & 6 & 17 \\ P_5 & 11 & 23 \\ P_6 & 5 & 1 \\ P_7 & 9 & 12 \\ P_8 & 7 & 33 \\ P_9 & 4 & 28 \\ P_{10} & 15 & 8 \\ P_{11} & 8 & 18 \end{array} \tag{4.2}$$

is depicted in Fig. 4.2, which shows the coordinates of points.

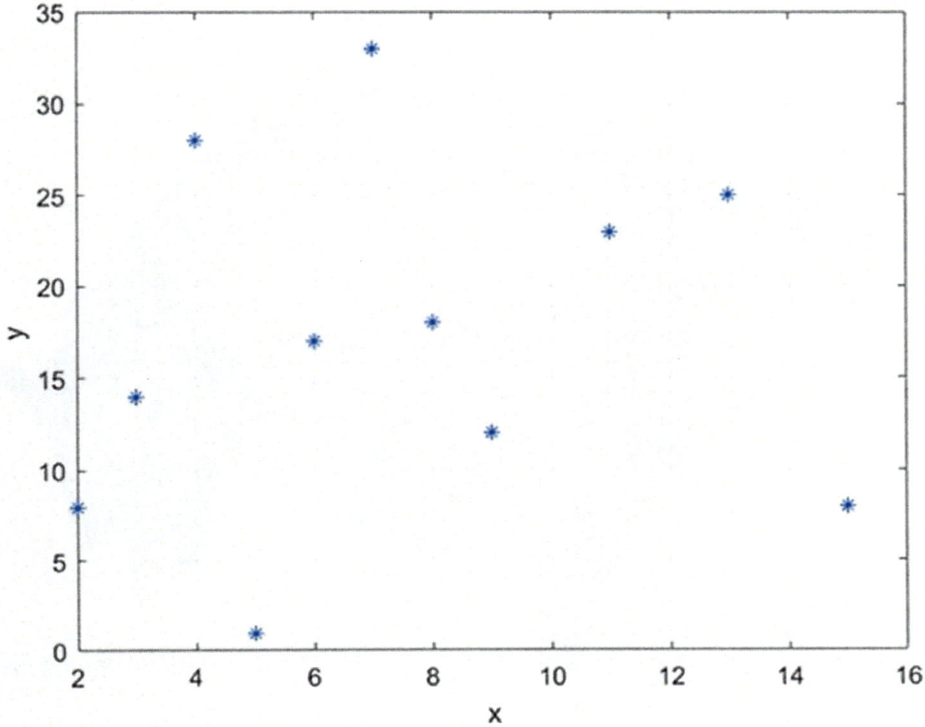

Fig. 4.2 11 point illustrative Example 4.2

Recall that the attraction force is defined as

$$w_i = \frac{1}{N-1} \sum_{j \neq i}^{N} e^{-\frac{\bar{d}_{ij}}{D} \alpha}$$

which, for the 10 point example, of TWC-Alpha is

Data	x	y	w_i
P_1	2	11	0.4084
P_2	2	13	0.4091
P_3	35	14	0.6900
P_4	28	14	0.7206
P_5	30	15	0.7221 .
P_6	34	10	0.6915
P_7	35	12	0.6921
P_8	33	16	0.6974
P_9	31	10	0.7127
P_{10}	28	12	0.7207

(4.3)

Table: Attraction Strength of Points of Table 4.1

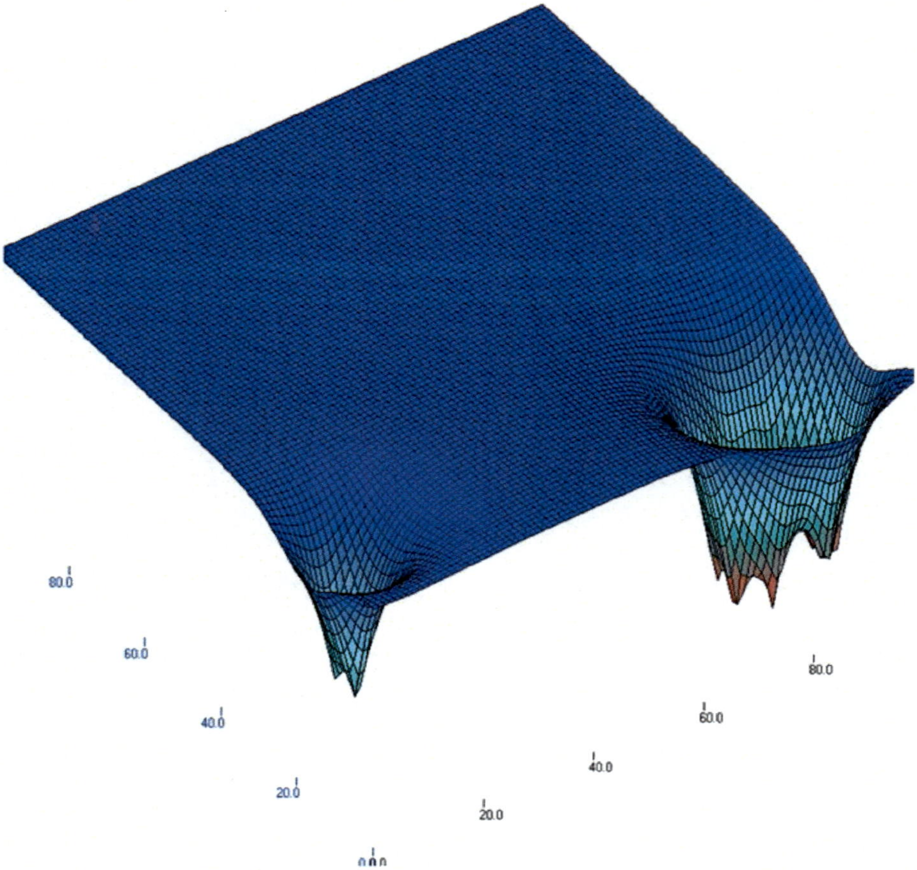

Fig. 4.3 Attraction strength field of 10 points

The depiction of the 10-point example as an inverted attractive force field is illustrated in Fig. 4.3 below.

The attractive force map is done in a similar way as that for the 11 point example.

4.2 TWC Maps

This section outlines the aspects that contribute to formation of the four (TWC—Alpha, TWC—Beta, TWC—Gamma, and TWC—Theta) of the five different TWC algorithms omitting the TWC-Iota analysis, and their associated scalar fields, their maps, in greater detail with respect to the two example data sets of 10 and 11 points. As we have already mentioned, TWC is a whole theory composed of several elements. Each of the elements is indispensable for a general understanding of the process.

Figure 3.25, depicted the five TWC algorithms four of which we discuss next with respect to the 10 and 11 point simple examples.

Remark 4.1 The MATLAB code used to generate the maps and graphs associated with this section can be found in the Appendix.

4.2.1 TWC-Alpha

The value of α at which free energy attains its maximum value corresponds to the α where the points of the distribution manifest the maximum level of attraction strength among them. Figure 4.1 shows the movement of the alpha value from the center of mass to the position of maximum attraction for the Example 4.1 defined over $U = [0, 60] \times [0, 36] \subset \mathbb{R}^2$. Intuitively, the movement makes perfect sense.

As can be seen from Fig. 4.4, the algorithm is moving the alpha point toward a location where it can "hear" sounds from each of the data points in the most effective way. Since this is a "patterned" set of points, the movement is also "patterned". In particular, there is symmetry (top/bottom) so the movement is along a straight line.

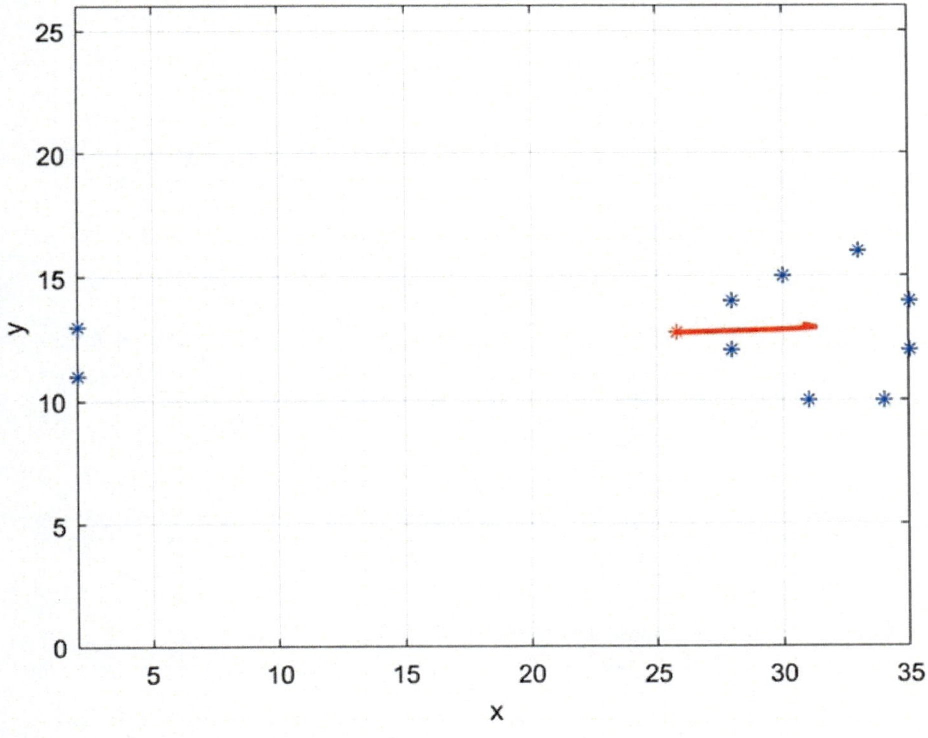

Fig. 4.4 TWC alpha point movement Example 4.1

The alpha point location is where the free energy is maximal and perhaps one can detect that there is a "turning back" in Fig. 4.4, it is faint. The alpha point corresponds to where the path turns back. For Example 4.2, Fig. 4.5 depicts the movement that identifies the alpha point. Since the points are random in the grid, the maximum alpha where it can "hear" all of the sounds coming from the data points most clearly would be close to the center of mass of the given data points.

A zoom of the subtle movement from the center of mass to the alpha point is depicted in Fig. 4.6. There is only a small change from the center of mass since the points are more or less randomly located on the grid. That is, the center of mass is relatively near the location from which sounds emanating from the 11 points can all be heard. The alpha point actually moves a slight distance and stops as soon as the free energy decreases (also corresponds to the minimal entropy), which is located at the first turning point. We let the algorithm continue for illustrative purposes, that is, so the location which is almost at the center of gravity (right endpoint) can be better seen.

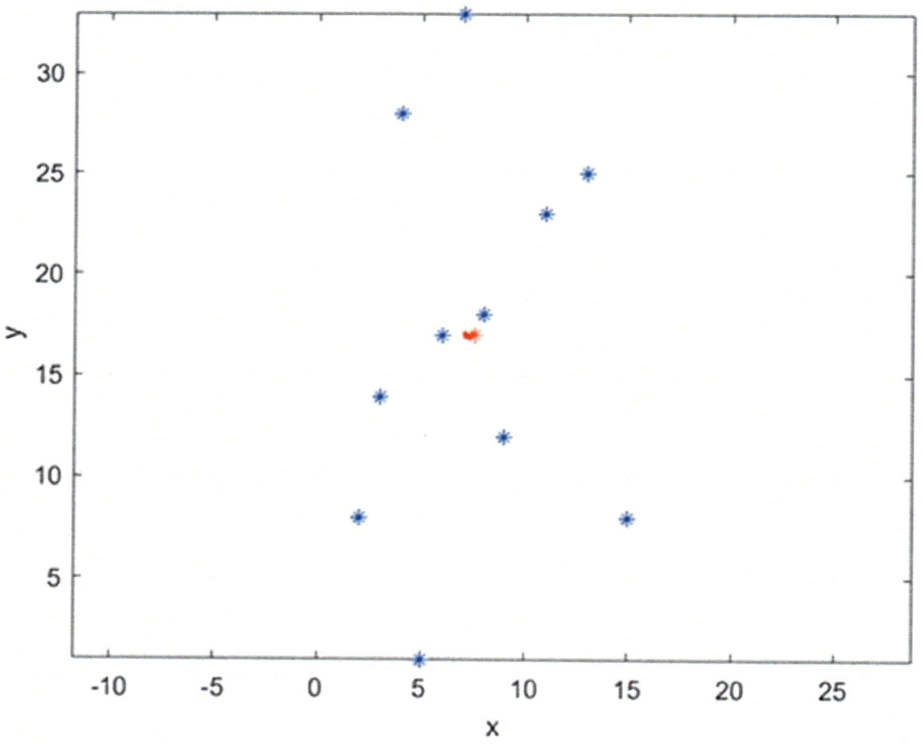

Fig. 4.5 TWC alpha point movement Example 4.2

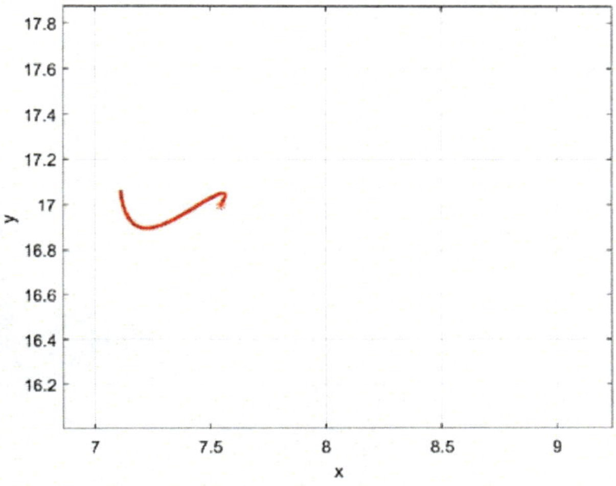

Fig. 4.6 Zoom of the movement of the alpha point—Example 4.2

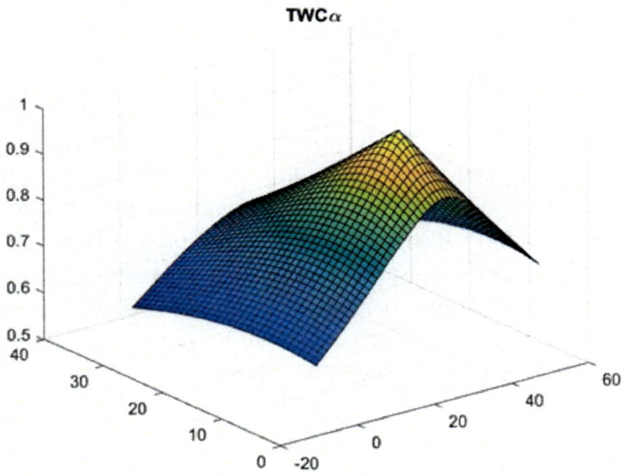

Fig. 4.7 TWC-Alpha Example 4.1

The map (the "heat" map) associated with the alpha point for Examples 4.1 and 4.2 are presented next. For Example 4.1, since there is a relatively patterned set of points (symmetric about the a horizontal line) with the location of the "sounds" more or less circular with two outliers, we see a relatively but somewhat patterned distribution a little off center of the circle of points as can be verified in Fig. 4.7.

TWC-Alpah for Example 4.2, since it consists of a random set of point, has its alpha point tightly located, see Fig. 4.8. That is, there is more of a single location from which sounds can be heard coming from the points of the given data points but the center of mass is roughly where all sounds are best heard.

Fig. **4.8** TWC-Apha Example 4.2

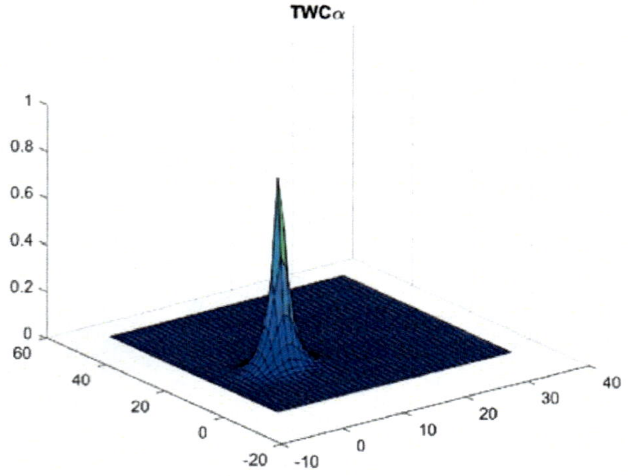

4.2.2 *TWC-Beta*

The TWC-Beta maps were generated in several formats. Recall that TWC-Beta is associated with where points in the short term future will locate themselves. It is the first prediction of the movement of the data points in the short term. Notice that the TWC-Beta algorithm is beginning to "hedge" its bets. Using the analogy of predator/prey, if the configuration of prey is Example 4.1, the predator would locate itself at the alpha point depicted by Fig. 4.7. As time goes on, the predator might consider the outliers and spend some of its energy according to the TWC-Beta map depicted by Fig. 4.9.

Another view of the TWC-Beta is Fig. 4.10.

Fig. 4.9 TWC-Beta Example 4.1

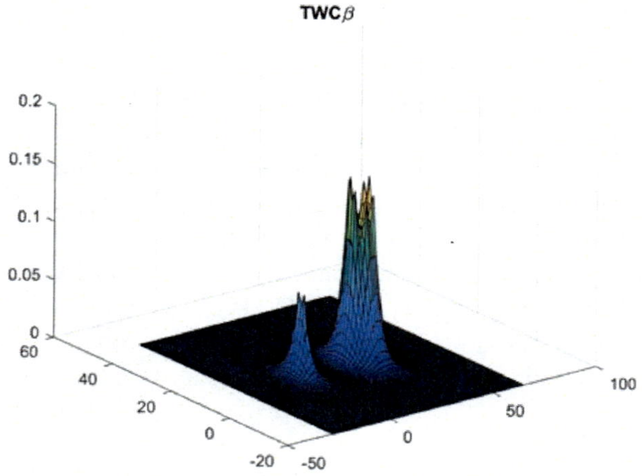

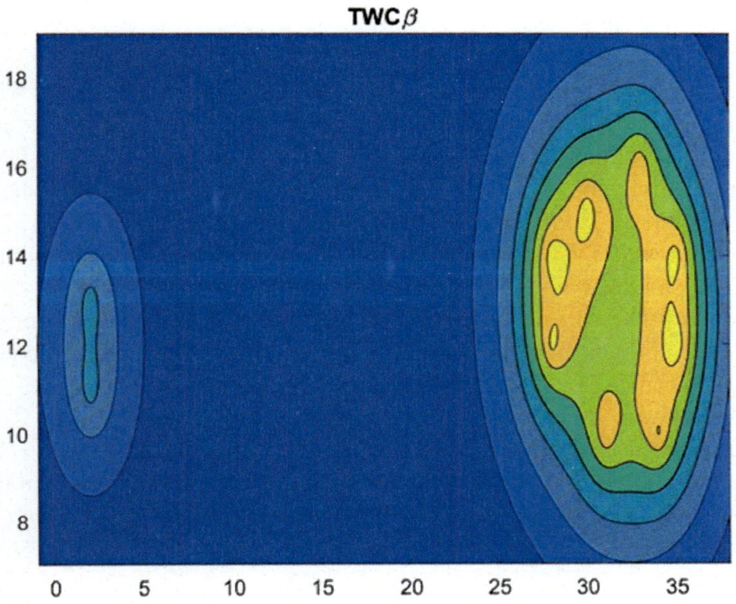

Fig. 4.10 TWC-Bet TWC-Beta Example 4.1 View 2

Fig. 4.11 TWC-Beta Example 4.2

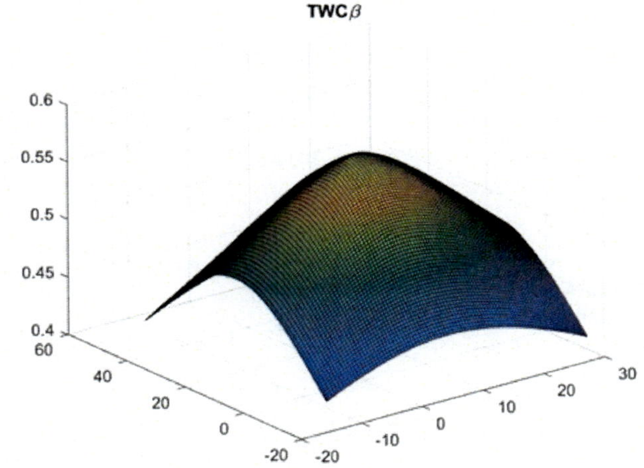

The TWC-Beta map for the Example 4.2 data, Fig. 4.11, being more of a random distribution of points, does not change much compared with its TWC-Alpha map. This is to be expected since a short term future configuration would be roughly the same as the past configuration.

4.2.3 TWC-Gamma

The TWC-Gamma further projects the dynamics of data set. Compared with TWC-Apha and TWC-Beta the configuration it seems to be focusing on the set of the original individual data points. Figure 4.12 is the illustration of TWC-Gamma for Example 4.1. In the predator/prey analogy, the predator focuses, in the near future, on each of the prey. Another view of this configuration is Fig. 4.13.

Associated with TWC-Gamma are associated, with respect to the predator/prey analogy, the path over which the predator would follow to attack the prey. It is clear, from Fig. 4.14 that the paths would be what is expected from such a configuration.

The TWC-Gamma map for Example 4.2, Fig. 4.15, shows a similar concentration of "energy" around the original set of individual data points.

Another view of this configuration is given by Fig. 4.16.

The associated gamma path for Example 4.2 is illustrate in Fig. 4.17. Note that paths are most strongly in the vertical direction rather than the horizontal indicating paths that a predator most likely might follow in pursuit of prey, or pathways a disease outbreak might follow.

4.2.4 TWC-Theta

TWC-Theta produces a longer term projection of the evolution of the dynamics of the data set. For Example 4.1, the configuration represented as a graph looks at how the points organize themselves over a longer period of time (see Fig. 4.18). The links between the points are as might be expected from this data set.

The undirected graph of weights for Example 4.1 is depicted by Fig. 4.19.

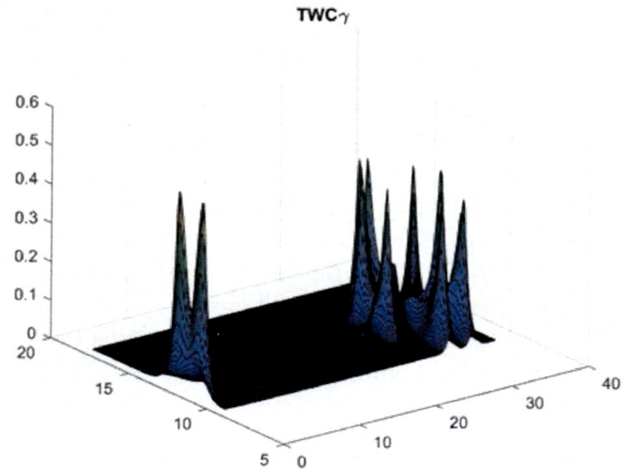

Fig. 4.12 TWC-Gamma Example 4.1

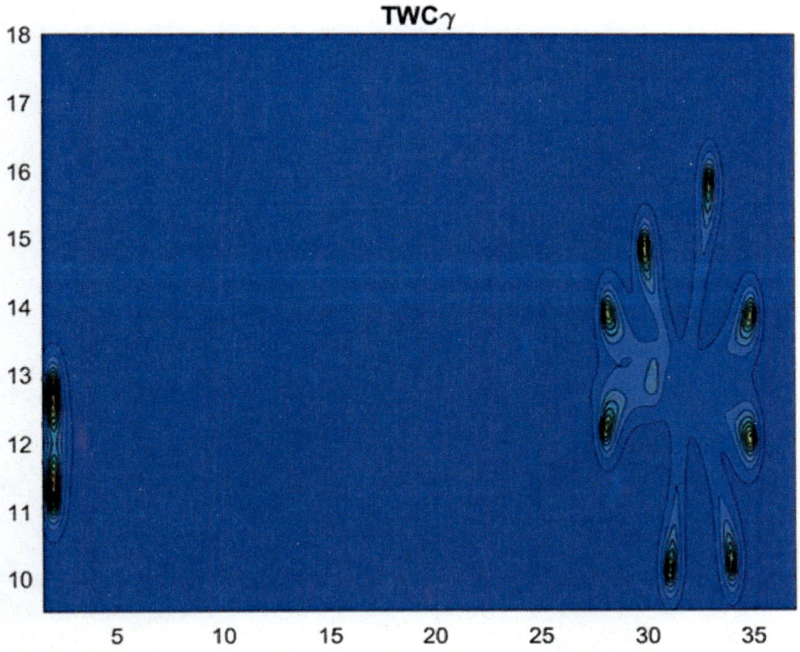

Fig. 4.13 TWC-Gamma Example 4.1 View 2

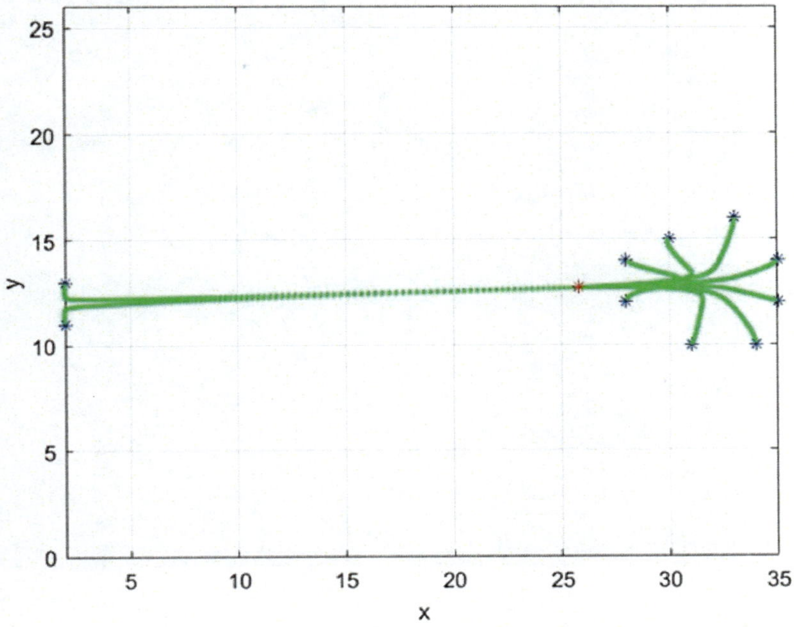

Fig. 4.14 TWC-Gamma paths Example 4.1

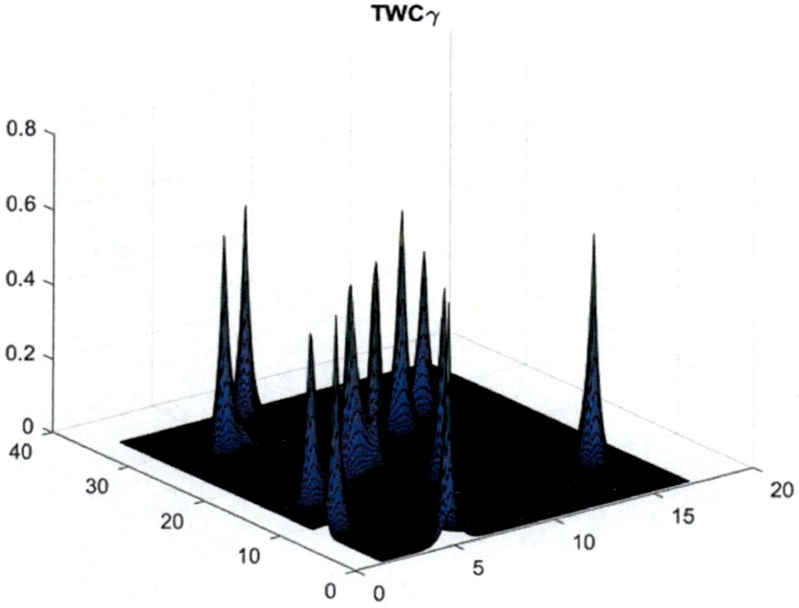

Fig. 4.15 TWC-Gamma Example 4.2

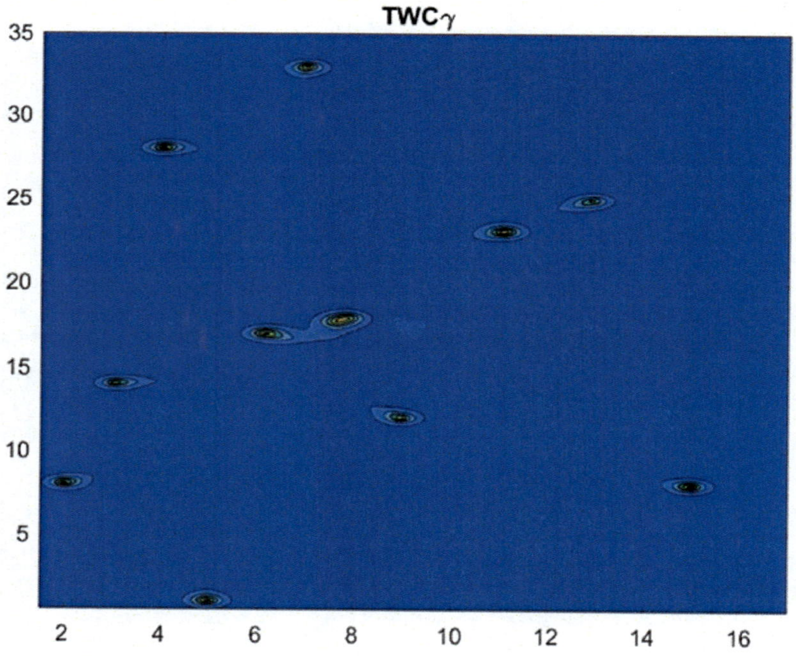

Fig. 4.16 TWC-Gamma Example 4.2 view 2

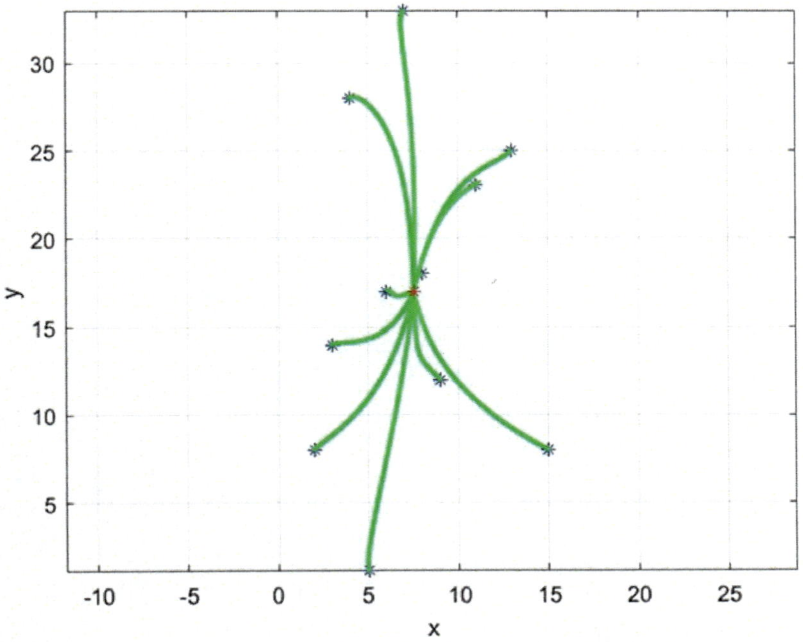

Fig. 4.17 TWC-Gamma path Example 4.2

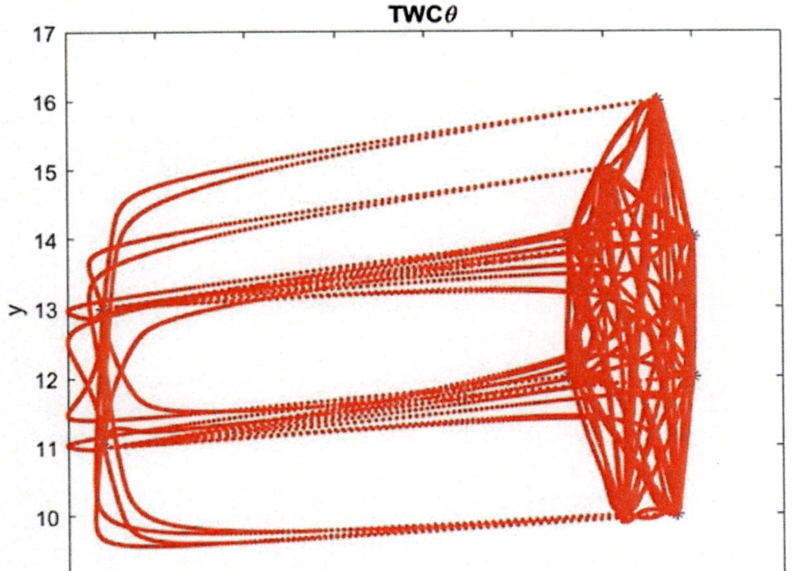

Fig. 4.18 TWC-Theta Example 4.1

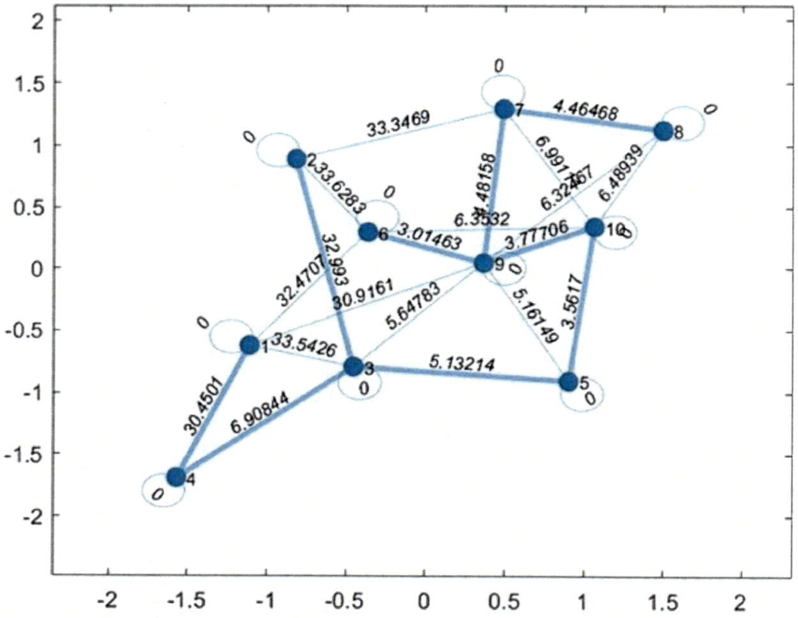

Fig. 4.19 TWC-THETA MST graph of weights Example 4.1s

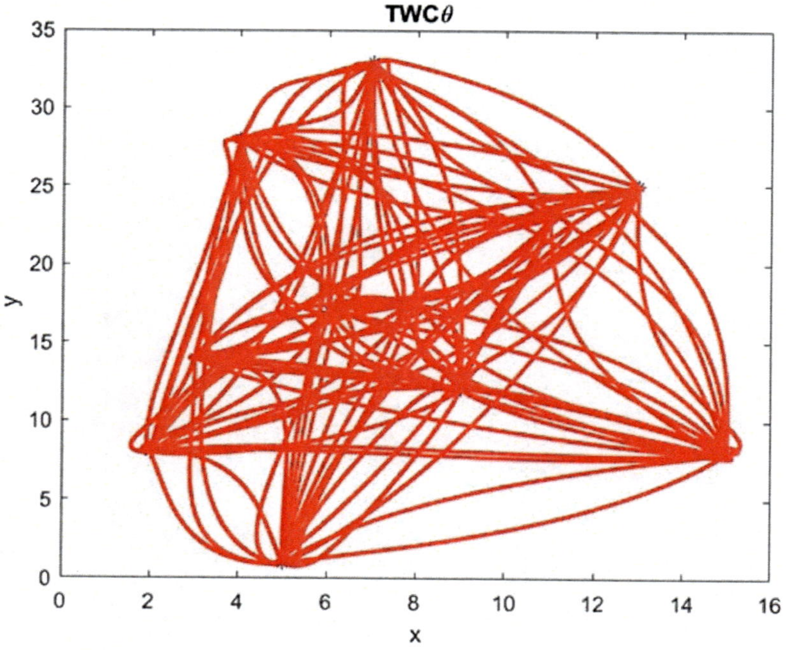

Fig. 4.20 TWC-Theta Example 4.2

For Example 4.2, the pattern is roughly a connected graph of the given data points as they are more or less randomly distributed. The graph associated with TWC-Theta for Example 4.2 is illustrated by Fig. 4.20. The MST weighted undirected graph for Example 4.2 are similar to Example 4.1 but quite "busy", so it is not included.

4.3 Summary

The two simple example illustrate the way that the TWC algorithms work on these two data sets. The evolution of the points from the point of view of free energy and entropy can be seen to be compatible with how these two principles work.

Chapter 5
Advanced TWC Topics

The previous chapters mentioned two types of analyses, the inclusion of semantic data associated with spatial data and the inclusion of local effects when they are significant so that these are not "washed out" by the global effects. This chapter begins with the development of how TWC balances local effects with the global ones. Then, this is followed by the discussion of how to incorporate what we term "semantic" data associated with geographic (location) data.

5.1 TWC Windowing

TWC-Widowing algorithm proceeds as follows. For each point of the data set selected a priori, neighboring points of the data set define its window. The size of the window can be small or large. The five TWC algorithms can then be applied in each window. After all the analyses (TWC—Alpha, Beta, Gamma, Theta, Iota) on each window is complete, all TWC maps are superimposed as a single map. The window maps, which depend on their size, might have large or small number of overlaps. This will make the superimposed TWC a network. Basically, each window around a point can be interpreted as a local TWC analysis between that point with its associated group of points and the superimposed TWC network over all data points that are its global analysis.

TWC-Windowing selects an *a priori* integer $K < N$ for segmenting (windowing) the N data points. The number of points in each window is $P = N/K$. In practice $K << N$. For each of the points, the P closest points will be found. For each data point there is a window. For the N data points there are P windows with each window having K points. Then, the five TWC algorithms can be applied on each window. As mentioned, all maps are superimposed and normalized as the final map. If $K = 1$, $P = N/K = N$, then TWC-Windowing becomes TWC-Original.

M. Buscema et al., *The Topological Weighted Centroid: A New Vision of Geographic Profiling*, Studies in Computational Intelligence 1095,
https://doi.org/10.1007/978-3-031-28901-9_5

5.2 Semantic TWC

The TWC suite of algorithms can also incorporate properties associated with the data set. For example, a data point might not only register its occurrence, but some added characteristic such as the severity of the event or the number of days (on average) that the event lasted. This is then put together with the location information as described next.

5.2.1 Introduction

The spatial dimension is often a key feature to understand the structure of a phenomena. In some cases, this is relatively obvious, as phenomena themselves are basically defined in spatial terms. Think, for example, of diffusion processes such as the acceptance of solar panels to be put up on the property. In some other cases, however, such a dimension is not obviously analyzed geographically. One might also consider a comparative analysis of different socioeconomic events occurring in a geographical region that are not necessarily part of the coordinate information. There is a vast array of alternative, statistically based approaches that have been developed to deal with the spatial character of phenomena in different fields, from physics, to biology, to economics and geography, and many more. However, an aspect that has been somewhat overlooked so far, is that of the semantic dimension of space, that is, interpreting the topological or metric dimension of space as conveying an intrinsic meaning that may have a substantial bearing on the interpretation of the underlying phenomena. That is to say, whether or not we are considering a physical space or some sort of abstract, representational space, it is often useful to consider that the spatial dimension may carry relational information as to why certain entities "stay together" in a given environment. This may add substantially to the understanding of the respective phenomena.

It is important to understand the meaning that we give to the term "relational" in this context. We are claiming that, to truly understand a certain phenomenon, we have to consider its variables as "negotiating" their position within the state space according to a semantics defined in terms of relative proximity. In other words, the physical entities produced by a given phenomenon situate themselves in the state space as if they were aware to some degree of their relative position, and would adjust to each other in appropriate ways as if they were abiding by a sort of implicit grammar of the phenomenon. We can unscramble and interpret such negotiation by suitably remapping the space in such a way to give to proximity its most expressive meaning, that is, by saturating as much as possible the semantics, compared with a context-insensitive, null semantics of self-referential spatial coordinates. Insofar as the semantic aspects of the phenomenon relate to the spatial characteristics only, we are basically reasoning in terms of a syntax of space, that is, the internal rules by which spatial features combine appropriately to generate proper structures of meaning. This is the most basic level of a relational spatial analysis. If, in addition, the phenomenon presents some characteristic dimensions of a nonspatial nature, we can speak of a full semantic. In this case, we will have to develop a more articulated approach. In previous chapters a full-fledged approach to space syntax has been presented. In this chapter, we bring the methodology one step forward to develop an approach to space semantics.

Consider a data set of R-*records* and V-*variables*. It can always be seen as a set of R points in a V-dimensional space. We can then introduce some quantities to describe some semantic aspects of the data set that are related to the relative position of the R records in their V-dimensional space. By the same token, one may transpose the reasoning and regard the same quantities as illustrating some semantic aspects of the relative positions of the V variables in their R-dimensional space provided that such transposition is computationally feasible.

All the quantities that define our approach will be introduced considering a set of K points, called entities, in a 2-dimensional space. This can be extended to the 3-dimensional case in a straightforward manner. The classic TWC quantities that we had previously defined are the following.

1. TWC-Alpha Point: A point in the map defining the outbreak of the process in the past
2. TWC-Alpha Map: An area in the map defining the outbreak of the process in the past
3. TWC-Beta Map: A Scalar field defining the hot areas in the moment in which data were collected
4. TWC-Gamma Map: A Scalar field defining the hot areas where the process tends to go in a short time
5. TWC-Theta Map: A Scalar field defining the hot areas where the process tends to go in a midterm
6. TWC-Theta Trajectories: The hypothetical trajectories connecting the assigned points (entities)
7. TWC-Theta Markov Machine: A direct and weighted graph showing how all the entities are connected
8. TWC-Iota Points: The new points where the entities might go to
9. TWC-Iota Map: The main area(s) where the entire process seems to come from and/or go to

These quantities imply that each entity of the K points (entities) has the same features, apart from its position within the N-dimensional space. However, we can, additionally, introduce the possibility that each entity has different features other than its position, thereby causing more complex interactions with the others within the N-dimensional ambient space. To deal with this further complication, we will apply an algorithm, named auto-contractive map (Auto-CM, see [1]) to evaluate the relationships among the K nonhomogeneous entities, taking their specific qualitative features into account. Furthermore, we will propose a new method to combine our derived relationships among nonhomogeneous entities with the pre-existing geographical information.

5.2.2 Theory and Equations

Assigned data points can be associated with an attribute, a quality for each assigned point (entities) of the process is considered. In this case we transform data set given by Eq. (5.1) into a data set represented by Eq. (5.2). We begin with a usual data set of points,

$$Data = \left\{ \left\{ P\left(x_{i,t}, y_{i,t}\right) \right\}_{i=1}^{N} \right\}_{t=1}^{T}, \tag{5.1}$$

where, T is the number of the temporal data points and $x_{i,t}$ and $y_{i,t}$ are longitude and latitude of the point i at time step t. When each data-point is characterized by a set of specific features, Equation (5.1) is written as

$$Data = \left\{ \left\{ P\left(x_{i,t}, y_{i,t}, \left\{a_{i,k}\right\}_{k=1}^{M}\right) \right\}_{i=1}^{N} \right\}_{t=1}^{T}, \tag{5.2}$$

Moreover, recall that the metric considered among the entities is not the Euclidean distance, but the pseudo-distance, Eq. (5.3).

$$\overline{d}_{ij} = \frac{1}{N-2} \sum_{\substack{k \neq j \\ k \neq i}} d_{ik}^{EU} \tag{5.3}$$

$$d_{ik}^{EU} = \sqrt{\sum_{l}^{N} \left(x_{il} - x_{kl}\right)^2}$$

A specific unsupervised ANN, Auto Contractive Map (Auto-CM) (see [1]), is used to find out patterns in the semantic features among the assigned points. Auto-CM is trained in order to provide the semantic association among any pair of points. That is, after the training phase, the Auto-CM weight matrix is a tensors matrix that can be thought as defining how each point (entity) is globally related to all the others. If the data set presents N points, then Auto-CM, after the training phase, will generate a square and nonsymmetric matrix of $N \times N$ weights (W). The main develop of and associated equations of the Auto-CM learning process with semantic data can be found in the article [2].

The process of obtaining relationships among the attributes involves several steps. The first operation is to transform the Auto-CM weighted matrix (W) into a quasi-distance, denoted $d^{[CM]}$, to be defined subsequently. The second operation is to scale $d^{[CM]}$ according to the same magnitude of the Euclidean distance among the points (entities). The first operation is represented by Eqs. (5.4) and (5.5) given next.

$$MaxCMWeight = \max(W_{ij}) \tag{5.4}$$

$$d_{ij}^{[CM]} = MaxCMWeight - W_{ij} \tag{5.5}$$

The second operation is represented by Eqs. (5.6)–(5.12) given next.

$$MaxD(CM) = \max\left(d_{ij}^{[CM]}\right) \tag{5.6}$$

$$MinD(CM) = \min\left(d_{ij}^{[CM]}\right) \tag{5.7}$$

$$MaxD(EU) = \max\left(d_{ij}^{[EU]}\right) \tag{5.8}$$

$$MinD(EU) = \min\left(d_{ij}^{[EU]}\right) \tag{5.9}$$

The scaling and offset factors are given as follows.

$$Scale = \frac{MaxD(EU) - MinD(EU)}{MaxD(CM) - MinD(CM)} \tag{5.10}$$

$$Offset = \frac{(MaxD(CM) \cdot MinD(EU) - MinD(CM) \cdot MaxD(EU)}{MaxD(CM) - MinD(CM)} \tag{5.11}$$

The new Auto-CM distance among points from a Semantic TWC point of view is given by:

$$d_{ij}^{*[CM]} = Scale \cdot d_{ij}^{[CM]} + Offset. \tag{5.12}$$

Now we need to analyze how the TWC equations change when we want to include the semantic features in every calculation. The first task is to modify the strength of attraction (α) and the self-attraction strength (β) of each entity. The already presented attraction strength, Eqs. (5.13) and (5.14) will be transformed into attraction strength Eqs. (5.15) and (5.16). The previous attraction strength is changed from

$$w_i(\alpha_n) = \frac{1}{N-1} \sum_{j=1, j\neq i} e^{-\frac{\overline{d_{ij}}}{D}\alpha_n}; \tag{5.13}$$

$$w_i(\beta_n) = \frac{1}{N-1} \sum_{j=1, j\neq i} e^{-\frac{\overline{d_{ij}}}{D}\beta_n}; \tag{5.14}$$

to

$$w_i^{[CM]}(\alpha_n) = \frac{1}{N-1} \sum_{j=1, j\neq i} e^{-\frac{d_{ij}+d^{[CM]}}{D}\alpha_n}; \tag{5.15}$$

$$w_i^{[CM]}(\beta_n) = \frac{1}{N-1} \sum_{j=1, j\neq i} e^{-\frac{d_{ij}+d^{[CM]}}{D}\beta_n}. \tag{5.16}$$

A modification of the scalar field of the Alpha Map is the next step. Equation (5.17), measuring the Euclidean distance between each point of the grid and every PMP, will be changed into the Eqs. (5.18)–(5.21) as follows. That is,

$$m_{kj} = \sqrt{(x_k - TWC_x(\alpha_k))^2 \left(y_k - TWC_y(\alpha_k)\right)} \tag{5.17}$$

is changed for Semantic TWC according to the next set of equations where we will modify d_{kj} into a new m_{kj}.

$$d_{kj} = \sqrt{(x_k - TWC_x(\alpha_k))^2 \left(y_k - TWC_y(\alpha_k)\right)}, \tag{5.18}$$

is the geometric distance between a grid point and an PMP.

$$MeanD_i^{[CM]} = \frac{1}{N-1}\sum_n^N d_{in}^{*[CM]}, \tag{5.19}$$

is the average of semantic distance of each entity from the others.

$$MeanD_i^{[PSEUDO]} = \frac{1}{N-1}\sum_n^N \overline{d}_{in}, \tag{5.20}$$

is the average of pseudo-geographic distance of each entity from the others.

$$m_{kj} = \frac{\left(d_{kj}^{[EU]} \cdot \frac{1}{N}\sum_i^N MeanD_i^{[CM]}\right)}{\frac{1}{N}\sum_i^N MeanD_i^{[PSEUDO]}}, \tag{5.21}$$

is the semantic distance between a grid point and an alpha point. Equation (5.22) leaves the calculation the activation of each grid point of the alpha scalar field the same as before, because the distance from each grid point and any PMP is already a semantic distance, we have,

$$a_k = \frac{1}{N}\sum_{j=0}^{N-1} e^{-\frac{m_{kj}}{D}\beta^*}. \tag{5.22}$$

Equation (5.23) that is used for the calculating of the geographical distance of each grid point from each entity, in order to estimate the beta map, will be changed into Eqs. (5.24)–(5.27) using to the same logic as above.

$$n_{kj} = \sqrt{\left(x_k - x_j\right)^2 \left(y_k - y_j\right)} \tag{5.23}$$

is changed for Semantic TWC according to the next set of equations.

$$d_{kj}^{[EU]} = \sqrt{\left(x_k - x_j\right)^2 + \left(y_k - y_j\right)^2}, \tag{5.24}$$

is the geometric distance between a grid point and a PMP.

$$MeanD_j^{[CM]} = \frac{1}{N-1}\sum_n^N d_{jn}^{[CM]}, \tag{5.25}$$

is the average of the semantic distance of each entity from the others.

$$MeanD_j^{[PSEUDO]} = \frac{1}{N-1}\sum_n^N \overline{d}_{jn}, \tag{5.26}$$

is the average of pseudo-geographic distance of each entity from the others.

$$n_{kj} = \frac{d_{kj}^{[EU]} \cdot MeanD_j^{[CM]}}{MeanD_j^{[PSEUDO]}} \tag{5.27}$$

is the resulting semantic matrix. As in the previous case, Eq. (5.28) leaves the calculation of the activation of each grid point of the beta scalar field the same as before, because the distance from of each grid point form any beta point is already a semantic distance. Thus we have

$$b_k = \frac{1}{N}\sum_{j=1}^N w_j \cdot e^{-\frac{n_{kj}}{D}\beta^*}. \tag{5.28}$$

Equation (5.29) used for the estimation of the gamma trajectories will be changed into Eq. (5.30) as follows.

$$p_{ij}\left(\gamma_i\left(t\right)\right) = e^{-\frac{\overline{d}_{ij}}{D}\gamma_i(t)} \tag{5.29}$$

will be changed to

$$p_{ij}\left(\gamma_i\left(t\right)\right) = e^{-\frac{\overline{d}_{ij}+d_{ij}^{[CM]}}{D}\gamma_i(t)}. \tag{5.30}$$

The calculation of the gamma map modifications are more complex. In fact we need to change Eqs. (5.31)–(5.35) in the following way.

$$m_{k\gamma_i(t)} = \sqrt{(x_k - TWC_x(\gamma_i(t)))^2\left(y_k - TWC_y(\gamma_i(t))\right)} \tag{5.31}$$

will be change to, repeating itself with a different symbol since $m_{k\gamma_i(t)}$ will be changed,

$$d_{k\gamma_i(t)} = \sqrt{(x_k - TWC_x(\gamma_i(t)))^2\left(y_k - TWC_y(\gamma_i(t))\right)} \tag{5.32}$$

$$MeanD_i^{[CM]} = \frac{1}{N-1}\sum_{n}^{N} d_{in}^{[CM]}, \tag{5.33}$$

$$MeanD_i^{[PSEUDO]} = \frac{1}{N-1}\sum_{n}^{N} \overline{d}_{in}, \tag{5.34}$$

$$m_{k\gamma_i(t)} = \frac{\left(d_{k\gamma_i(t)}^{[EU]} \cdot \frac{1}{N}\sum_{i}^{N} MeanD_i^{[CM]}\right)}{\frac{1}{N}\sum_{i}^{N} MeanD_i^{[PSEUDO]}}, \tag{5.35}$$

As in the previous case, Eq. (5.36) leaves the calculation the activation of each grid point of the gamma scalar field the same as before,

$$c_k = \frac{1}{\sum_{i}^{N} Q_i}\sum_{i=1}^{N}\sum_{t}^{Q_i} e^{-\frac{m_{k\gamma_i(t)}}{D}\beta^*}. \tag{5.36}$$

The theta map, to become a semantic map, follows the same procedure, but assuming this time that each theta point has to belong to the nonlinear MST. Consequently, Eq. (5.37) is replaced by Eqs. (5.38)–(5.41). Recall that previously,

$$l_{k\theta_z(t)} = \sqrt{\left(x_k - DirectPath\_\theta_{z_x}(t)\right)^2 + \left(y_k - DirectPath\_\theta_{z_{ywhi}}(t)\right)^2} \tag{5.37}$$

which is the distance between a grid point k and each of point of the nonlinear MST $\theta_z(t)$. This is changed to

$$d_{k\theta_z(t)} = \sqrt{\left(x_k - DirectPath\_\theta_{z_x}(t)\right)^2 + \left(y_k - DirectPath\_\theta_{z_{ywhi}}(t)\right)^2} \tag{5.38}$$

$$MeanD_i^{[CM]} = \frac{1}{N-1}\sum_{n}^{N} d_{in}^{[CM]}, \tag{5.39}$$

$$MeanD_i^{[PSEUDO]} = \frac{1}{N-1}\sum_{n}^{N} \overline{d}_{in}, \tag{5.40}$$

$$l_{k\theta_z(t)} = \frac{\left(d_{k\theta_z(t)} \cdot \frac{1}{N}\sum_i^N Mean D_i^{[CM]}\right)}{\frac{1}{N}\sum_i^N Mean D_i^{[PSEUDO]}}, \tag{5.41}$$

Equation from TWC-Theta (restated as Eq. (5.42)) remains unaltered, and it is used to calculate the theta semantic activation of each grid point.

$$e_k = \frac{1}{\frac{N-1}{\sum_z P_z}} \sum_{z=1}^{N} \sum_t^{P_z} e^{-\frac{l_{k\theta_z(t)}}{D}\beta^*}. \tag{5.42}$$

The Iota calculation changes only when the Euclidean distance is estimated at the first recursive iteration. Equations (5.43) and (5.44) define the scaled Euclidean distance at the first iteration ($t = 0$), but Eq. (5.44) has to be replaced with Eq. (5.45), in order to consider also the semantic features provided by the Auto-CM processing.

$$D_{ij}^{[t=0]} = \sqrt{\left(x_i - x_j\right)^2 + \left(y_i - y_j\right)^2} \tag{5.43}$$

$$d_{ij}^{[t=0]} = f(D_{ij}^{[t=0]}) \tag{5.44}$$

where $f()$ is a linear scaling of D that generates a value in the interval [0, 1] as previously discussed. The Semantic TWC changes the distance in the following way.

$$d_{ij}^{[t=0]} = \frac{1}{2}\left(f(D_{ij}^{[t=0]}) + \left(1 - w_{ij}^{[CM]}\right)\right), 0 < w_{ij}^{*[CM]} \leq 1 \tag{5.45}$$

5.2.3 A Comparison Between Semantic and Syntactic TWC (50 Terrorist Attacks in Afghanistan in 2009)

Fifty terrorist attacks were recorded in Afghanistan until May 2009. The author of the data set here presented has to remain unknown, but it is related to the USA Army. For each attack the latitude and longitude was recorded, with some attributes characterizing the quality and the quantity of that attack such as the tribe name of the author of the attack and its quantitative profile. Figure 5.1 shows the number of attacks executed by each tribe, and Fig. 5.2 shows the number of the attacks in the different areas of the country (West, Mid-West, Center, Mid-East, East). Figure 5.3 shows the geography of the 50 attacks on the map of Afghanistan. The attributes showed in Table 5.2 was transformed into 86 Boolean variables. Consequently, the semantic data set was composed of 50 attacks, each one defined by 86 attributes,

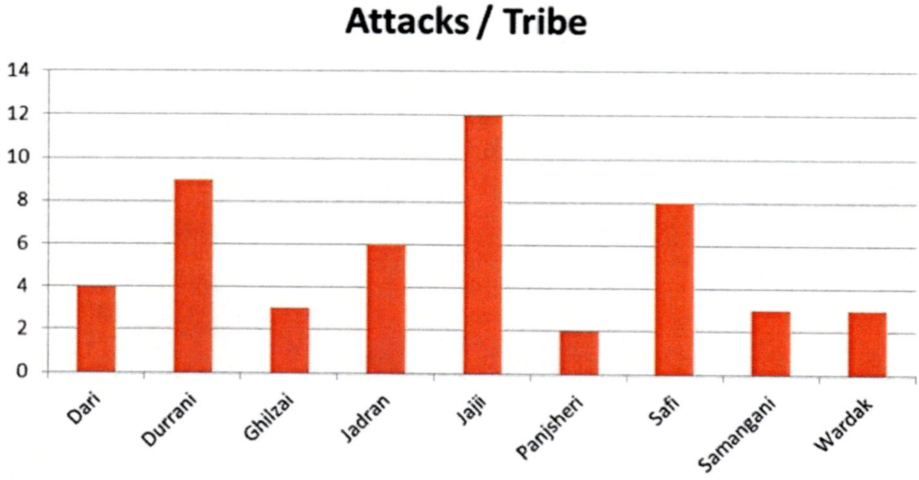

Fig. 5.1 Number of attacks executed by each tribe

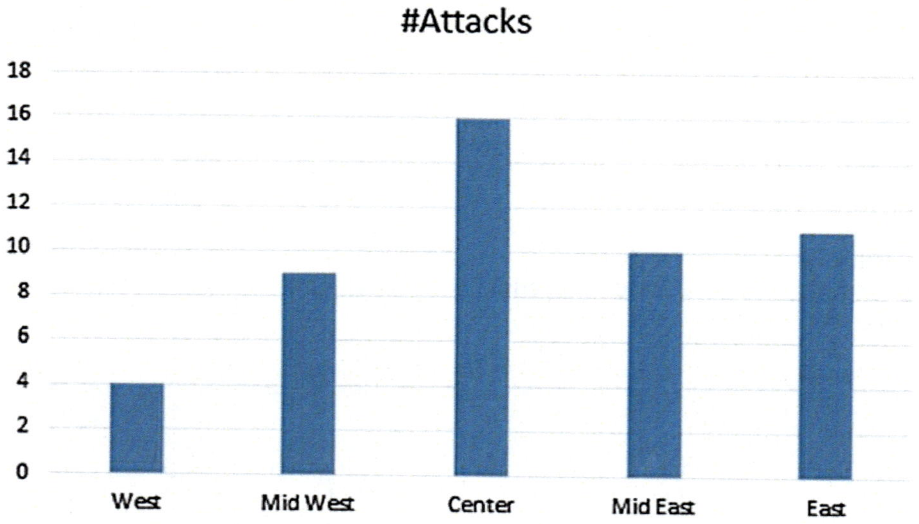

Fig. 5.2 Number of attacks in different areas of the country

while the syntactic data set was composed by the latitude and longitude of each one of the 50 attacks (see Eq. (5.46))

$$AttackData = \left\{ P\left(\mathrm{long}_i, \mathrm{lat}_i, \{militaryattribute_{ik}\}_{k=1}^{86}\right)\right\}_{i=1}^{50}. \tag{5.46}$$

Now we proceed to a systematic comparison between the syntactic and the semantic version of the TWC maps. In order to measure the difference between the syntactic and the semantic maps, two cost functions will be used: (1) The square correlation

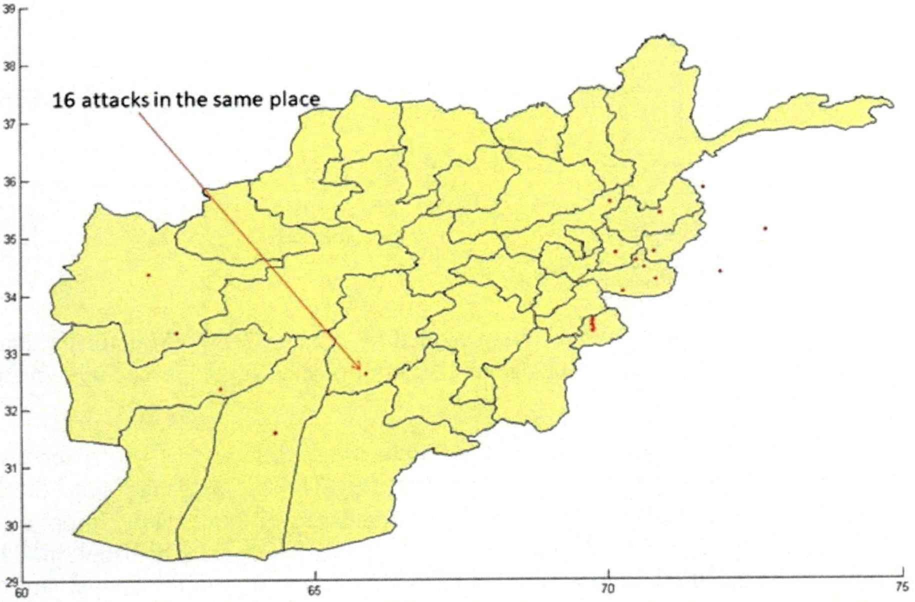

Fig. 5.3 Provinces of Afghanista

($R2$ below) and (2) The probabilistic density function similarity (PDF_Sim below), according to the Eqs. (5.47) and (5.48) respectively. The $R2$, Eq. (5.47), is a linear function based on the average and the variance. Consequently, it is not able to detect the topological differences among the maps, differences that are evident also from a visual point of view. So we will not take in account of the results of this measure.

PDF_Sim, Eq. (5.48), instead, is a nonlinear function able to detect better the topological differences among the maps. To this end, we have the following.

$$
\begin{aligned}
x_{ik} &= kth \text{ grid point of the } ith \text{ map} \\
x_{jk} &= kth \text{ grid point of the } jth \text{ map} \\
\overline{x}_i, \overline{x}_j &= \text{average of activation of all the grid points of the } ith \text{ and } jth \text{ map} \\
N &= \text{number of grid points of each map} \\
M &= \text{number of bin intervals of each map} \\
i, j &= \text{indices of each couple of maps to be compared} \\
b_{ik} &= \text{probability of the } kth \text{ bin of the } ith \text{ map} \\
b_{jk} &= \text{probability of the } kth \text{ bin of the } jth \text{ map}
\end{aligned}
$$

The associated equations are given next.

$$P_{ij}^{[R2]} = \left(\frac{\sum_{k=1}^{N} \left(x_{ik} - \overline{x}_j\right)\left(x_{jk} - \overline{x}_j\right)}{\sqrt{\sum_{k=1}^{N} \left(x_{ik} - \overline{x}_j\right)^2 \sum_{k=1}^{N} \left(x_{jk} - \overline{x}_j\right)^2}} \right), 0 \leq P_{ij}^{[R2]} \leq 1 \quad (5.47)$$

$$P_{ij}^{[PDF-Sim]} = \frac{1}{M} \sum_{k=1}^{M} \frac{b_{ik} b_{jk}}{\max\left\{b_{ik}, b_{jk}\right\}}, 0 \leq P_{ij}^{[PDF-Sim]} \leq 1 \quad (5.48)$$

Tables 5.1 and 5.2 show the results provided by $R2$ and PDF_Sim for the maps alpha, beta, gamma, theta and iota, generated by the syntactic version and by semantic version of TWC.

Alpha Maps: Fig. 5.4 shows the syntactic alpha map, while Fig. 5.5 shows the same alpha map processing with the semantic attributes. The two maps are quite similar ($PDF - Sim = 0.9270$). In other words, the processing of the semantic information does not change significantly the alpha map of this data points distribution. This happens because the outbreak of the attacks in both cases is in the area where the first 16 attacks also occurred. The area of the outbreak is in center of the Chora district in the province of Uruzgan, that was considered, at least up to 2017, as one of the main enclaves of terrorism in Afghanistan.

The Start Project (University of Maryland) stated: "09/30/2017: Assailants attacked Chora district in Uruzgan, Afghanistan. At least 31 assailants and 26 security personnel were killed and 14 assailants and three police officers were injured in the ensuing clash. No group claimed responsibility; however, sources attributed the attack to the Taliban". For this map, only the x and y coordinates of the 50 points were used. The alpha semantic map is depicted in Fig. 5.5.

Table 5.1 Similarity among maps according to R2

R2	Syntactic-semantic
Alpha-Map	0.998861427
Beta-Map	0.574693956
Gamma-Map	0.893119857
Theta-Map	0.929606288
Iota-Map	0.002284688

Table 5.2 Similarity among maps according to *PDF-Sim*

Maps	*PDF-Sim*
Alpha-Map	0.9270
Beta-Map	0.2404
Gamma-Map	0.7252
Theta-Map	0.6634
Iota-Map	0.5600

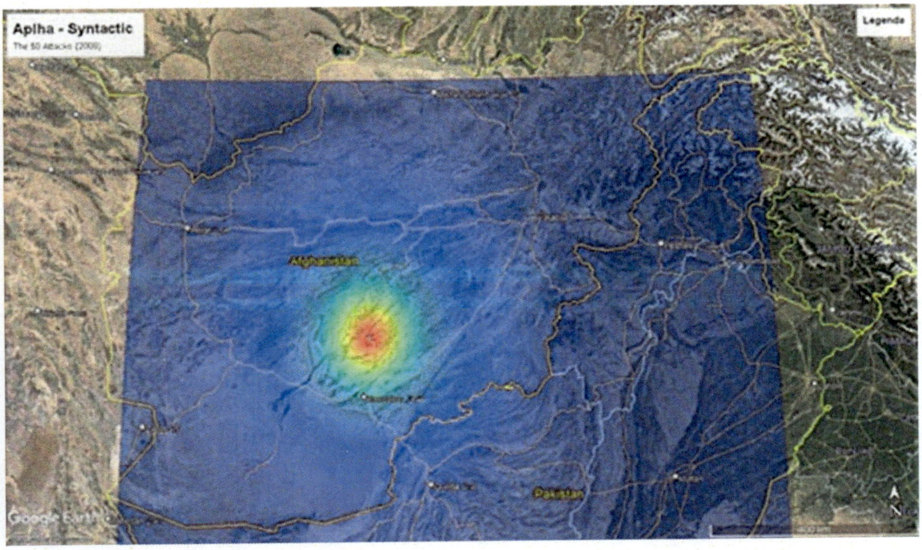

Fig. 5.4 Syntatic alpha map

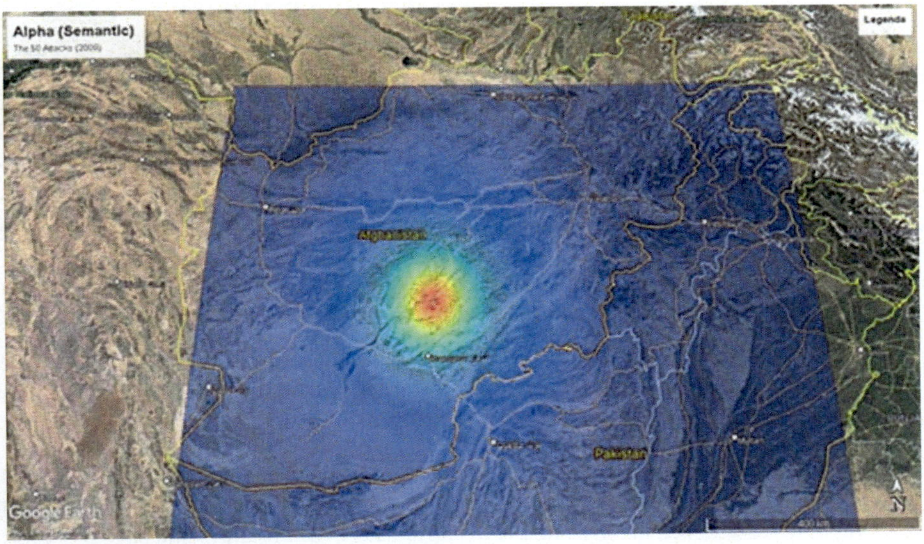

Fig. 5.5 In this figure alpha semantic map

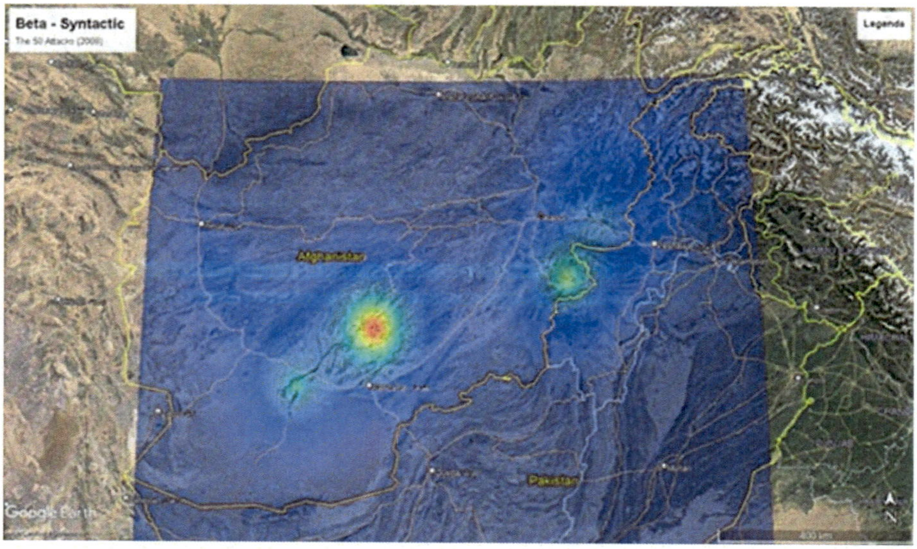

Fig. 5.6 Beta-syntatic

Beta Maps. Figures 5.6 and 5.7 show respectively syntactic version and the semantic version of the beta map. This time the two maps are very different ($PDF - Sim = 0.2404$). The semantic map is much more selective. It points out only two areas: the main one is again the Chora district in the province of Uruzgan, and the second one is around the city of Lashkar Gah (Helmand Province), a key zone for the Taliban activity. Moreover, the Start Project (University of Maryland) states the following, three months after the time in which these data points were analyzed with the TWC: "08/20/2009: On Thursday, more than 20 rockets fell in Lashkar Gah, the capital of Helmand province, in the south. One of the rockets landed near a line of voters at 0720, and killed a child, injured a civilian and damaged a polling station. No group has claimed responsibility, but it is believed that the attack is part of an effort of Taliban to disrupt the election".

Gamma Maps. Figures 5.8 and 5.9 show the gamma map of the syntactic and the semantic version of TWC. Also in this case the two maps are quite different ($PDF - Sim = 0.7252$). Because TWC gamma is intended to make a short term prediction of the probability distribution of the data points, the predictions of the two maps are completely divergent from the point of view of risk intensity, also if the hot areas are partially overlapped. According to the syntactic map the next step of the terrorist attacks should happen in north east of the country, while according to the semantic map new terror attacks would develop in south west area.

Theta Maps. Figures 5.10 and 5.11 show the theta map of the syntactic and the semantic version of TWC. The similarity of the two maps is very low ($PDF - Sim = 0.6634$). Syntactic theta map (Fig. 5.10) focuses the area on the east side of Kabul as the main area of future terror attacks (Pakistan boundary). The semantic

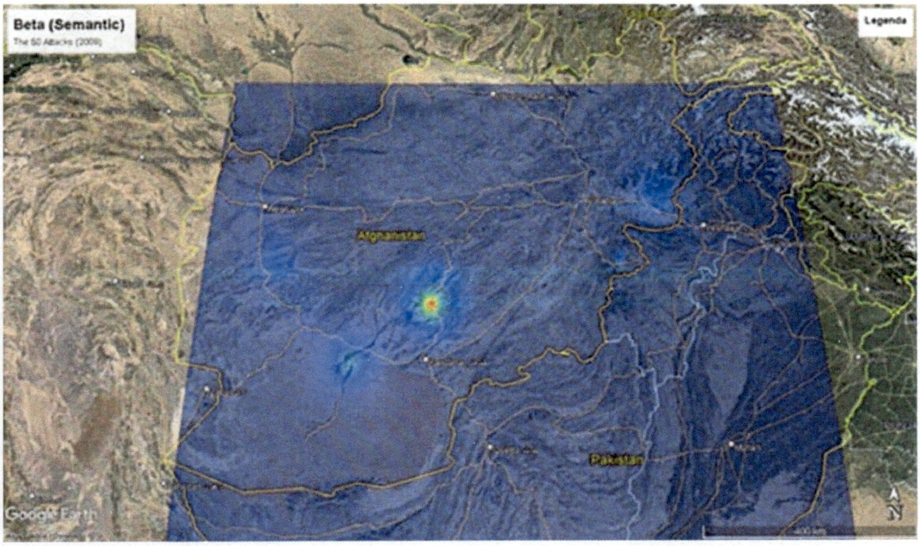

Fig. 5.7 Beta-semantic

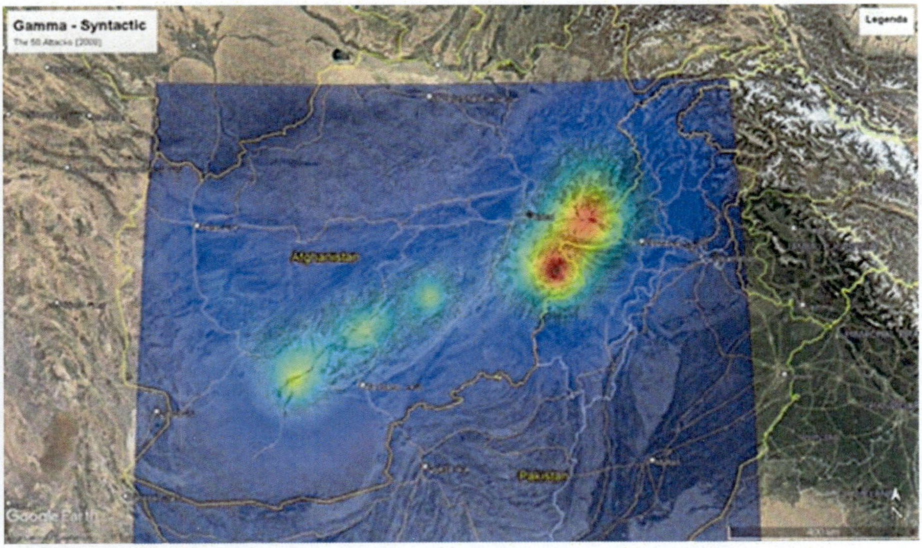

Fig. 5.8 Gamma-syntatic

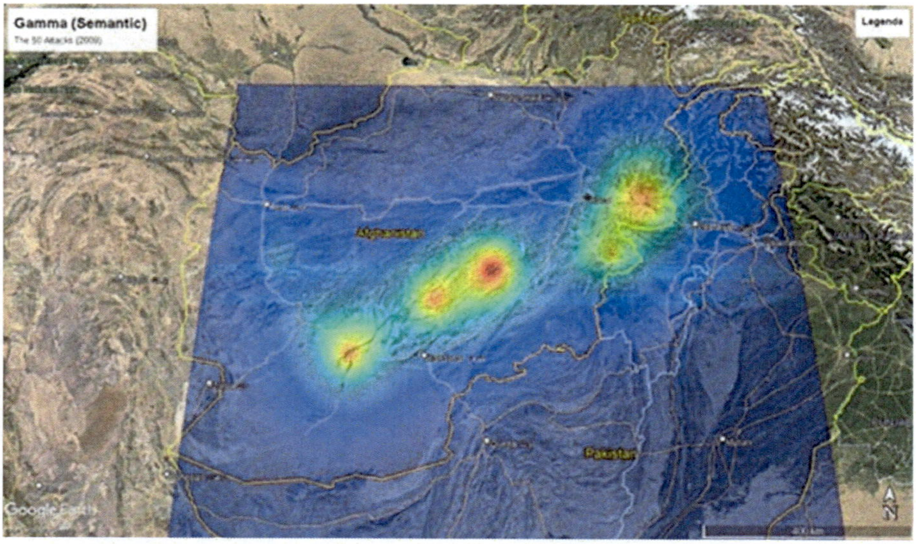

Fig. 5.9 Gamma-semantic

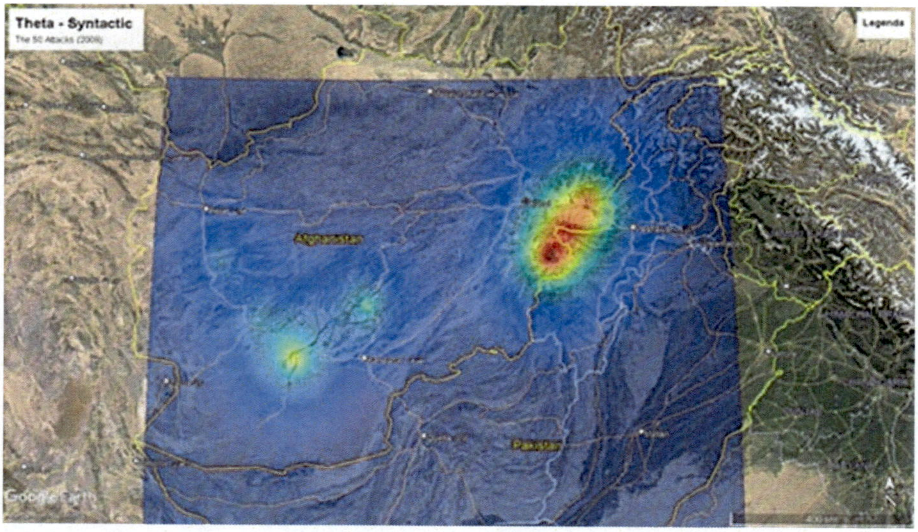

Fig. 5.10 Theta-syntatic

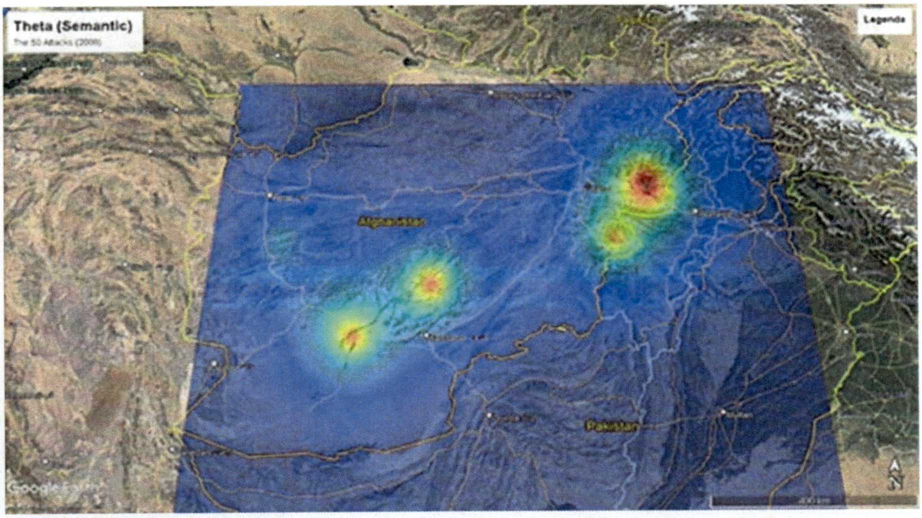

Fig. 5.11 Theta-semantic

Table 5.3 Terrorist attacks in Afghanistan in 2010, in 4 areas predicted by semantic TWC Theta in May 2009

Areas of attacks in 2010	Percentage %
Jalalabad area	7.05
Laskar Gah area	6.49
Chora district	3.90
Jaji district and Kabul area	9.28
Total	26.72

theta map, instead, points out four different areas as the main outbreaks of future attacks: the area around Jalalabad, the Chora district in the province of Uruzgan, the district of Jaji, and area around the city of Lashkar Gah (Helmand Province). In 2010 in Afghanistan 536 terrorist attacks were recorded, 27% of them (see Table 5.3) happened in the four areas predicted by the theta semantic map. Between 2010 and 2011 the number of attacks became 949, more than 31% of them in the provinces detected by TWC theta semantic, using only 50 attacks that occurred up to May 2009 (see Table 5.4).

Note that of the attacks noted in Table 5.4, 5 provinces were predicted by semantic TWC—Theta in May of 2009.

Iota Maps. Iota Map is generated by the TWC—Beta processing (Meta-Beta) of the new points (virtual points or vanishing points) projected by the Iota algorithm. Semantic and Syntactic version of the Iota Maps are quite different ($PDF - Sim = 0.5600$). Figures 5.12 and 5.13 show both the maps.

Table 5.4 Terrorist attacks 2010–2011

Terrorist attacks 2010–2011	
Province	Percentage %
Uruzgan	2.85
Helmand	9.48
Kabul	6.43
Jalalabad	7.06
Khost	5.48
Total	31.30

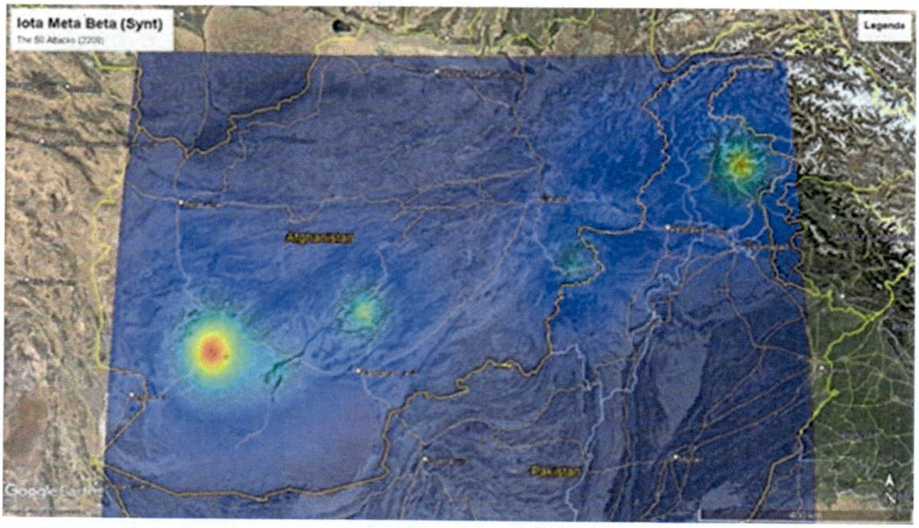

Fig. 5.12 Iota-syntatic

It is surprising that the syntactic iota map shows a hot area in Pakistan, which is out of the convex hull of the assigned 50 data points. This area is 92 km away from Abbottabad, where Bin Laden was hidden and then killed by a special USA Army mission two years after (Fig. 5.14).

It is also surprising what the semantic version of iota map shows. Figure 5.15 shows the terror attacks in Afghanistan from 2010 to 2014 (5671 red dots), overlapped onto the hot map of semantic TWC—Iota generated in 2009 using only 50 attacks. It is evident that the risk areas (red and/or yellow color) of the hot map occurred in the places where in the following years the most of the attacks happened, with two exceptions (red circles). It is worth analyzing these exceptions. The first exception of two Pakistan areas as risk zone for terrorism in Afghanistan (Fig. 5.16).

Remark 5.1 According to *CSIS*. PAKISTAN: VIOLENCE *Versus* STABILITY A National Net Assessment (May 5 2011):"A 2006 *UNODC* investigation found no

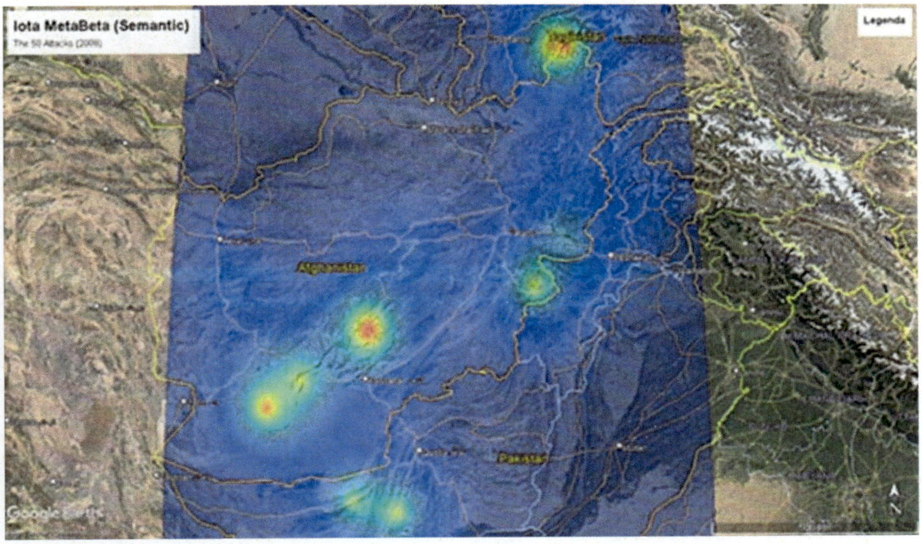

Fig. 5.13 Iota-semantic

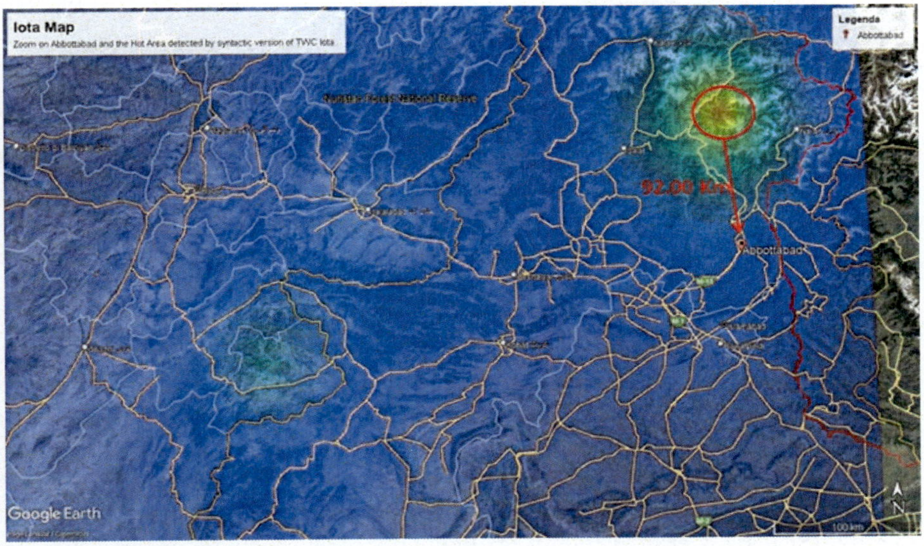

Fig. 5.14 Iota-map—Headquarters of Bin Laden

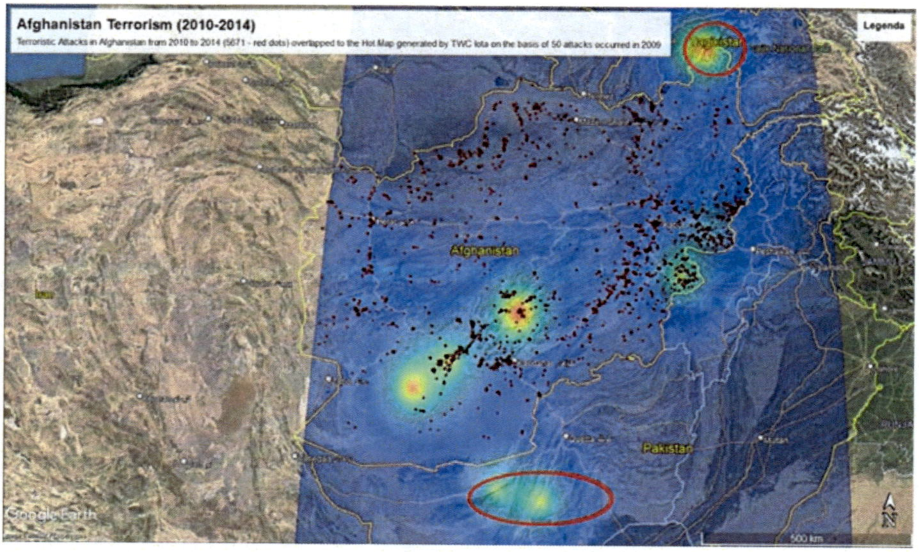

Fig. 5.15 Terrorist attacks 2010–2014

border posts along the long Chagai-Nimroz border, leaving open 300 km. 457 Frontier Corps sources also claimed that today there are only two posts along in the Chagai and Nushki districts, two districts with the longest borders with Afghanistan. As a result, the infiltration into Helmand, Kandahar, Nowshera and Shorawak is relatively easy, and it is believed that Afghan fighters sometimes use hospitals in the cross border town of Dalbandin in Chagai. The capital of Balochistan, Quetta, is a war zone, populated by a lethal mix of extremist nationalists, political separatists, religious fanatics, narcotics and weapons smugglers, as well as international terrorists and foreign intelligence agencies. 460 Sectarian violence and ethnic targeted killings are an almost daily occurrence as are attacks against government forces and institutions. Quetta has been a historical staging ground for war in Afghanistan, and is believed to be the command node for the Quetta Shura Taliban, including Mullah Omar. Senior leaders such as now deceased Mullah Dadullah openly maintained residences and hosted parties, including a September 2003 family wedding, at which JUI leaders, military officers and government officials were prominent guests. It is often suggested that the ISI plays a critical role in facilitating Taliban sanctuary and logistics, but a report by analyst Matt Waldman goes further and suggests that ISI observers are actually represented at the meetings of the Quetta Shura as observers."

The second exception is the north border between Afghanistan and Tajikistan.

Remark 5.2 From Lawrence P. Markowitz, Rowan University, State Responses To The Terrorism-Trafficking Nexus In Tajikistan: Short Term Advances, Long Term Risks, (216): "Tajikistan has become the most prominent destination for opium trafficked from Afghanistan. As illustrated in Fig. 5.1 (Renamed 5.17), most of the

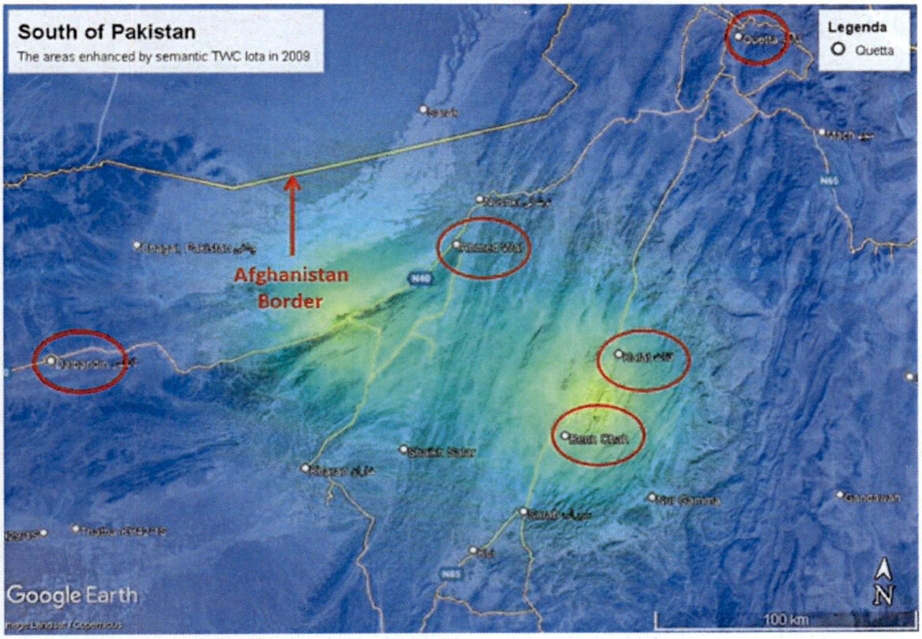

Fig. 5.16 South Pakistan

drug trafficking in Tajikistan pass through the densely populated Hatlon Province which borders Afghanistan's provinces of Badakhshan, Takhar, and Kunduz. Opium poppy cultivation has been ongoing in Afghanistan's Badakhstan Province, which also serves as the central point for drug trafficking networks in the region. Both Takhar and Kunduz serve as transit points for opium trafficking from other Afghan provinces [...]One indication of continued drug trafficking is the construction boom in Dushanbe over the past decade, which several interviewees viewed as a means of laundering money from narcotics sales." See https://ipsr.ku.edu/trafficking/pdf/PolicyBrief_Tajikistan.pdf.

Figure 5.17 shows a zoom of the area detected by TWC—Iota and Fig. 5.18 (the figure in the above quotation) represents the amount of drug trafficking between Afghanistan and Tajikistan. It is easy to understand that TWC—Iota has detected a risk area (or more properly a vanishing area), linked to terrorism, 6 years before Fig. 5.18 was published. In Fig. 5.18, indeed, the iota hot area is depicted as a known area dedicated to drug trafficking and terrorism. Note that the distance between the town of Dushanbe, the main center of the drug trafficking organizations, and the iota hot area is only 130 km away.

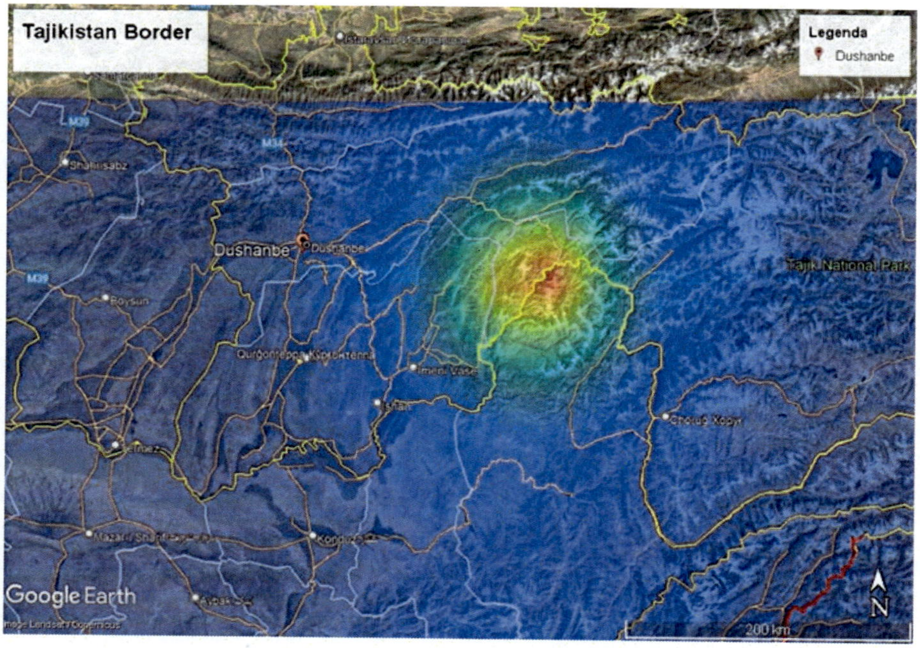

Fig. 5.17 North Pakistan

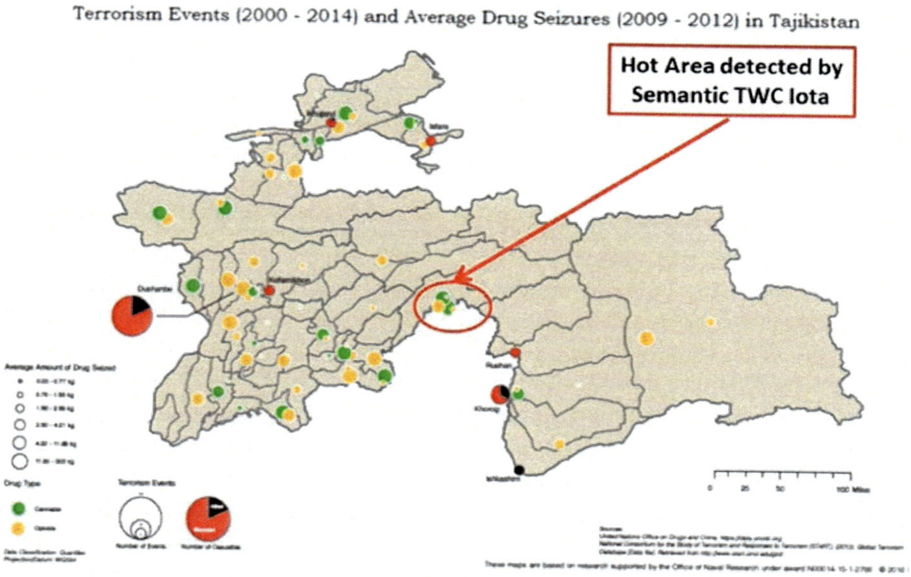

Fig. 5.18 North Pakistan

5.2.4 *Summary*

The semantic version of TWC produces different information from the syntactic version within the same interpretative framework. Both the types of information, however, have been shown to be useful in detecting hidden relationships embedded in the data points. Semantic TWC has shown to be more informative in most of the applications. This is especially true when the attributes qualifying all the data points are meaningful and representative.

References

1. M. Buscema, W. Lodwick, M. Breda, G. Massini, F. Newman, M.A. Zeydabadi, *Artificial Adaptive Systems Using Auto Contractive Maps: Theory, Applications and Extensions* (Springer, 2018). https://doi.org/10.1007/978-3-319-75049-1
2. M. Buscema, M. Breda, E. Grossi, L. Catzola, P.L. Sacco, Semantics of point spaces through the topological weighted centroid and other mathematical quantities: theory and applications, Chapter 4, pp 75–139, in *Data Mining Applications Using Artificial Adaptive Systems*. ed. by W.J. Tastle (Springer Science+Business Media, New York, 2013)

Chapter 6
TWC Applications

This section contains a few of the many applications of TWC that have actually been implemented either at the request of a person or as a project from an agency or as a part of a journal publication. A few have been done as a scientific experiment, one was part of a thesis. There are very many more applications that were successfully run but are not presented in this section some of which are found in the bibliography. The applications found in this section represent the variety of types of problems that have been solved. TWC began when we asked the following questions: (1) If a series of events occur in time and physical space, are these events connected? (2) Is there an underlying process that generates what is manifest in time/space? TWC algorithms, as can be seen by the applications, answers these questions in the affirmative.

The example applications that we present are similar to the following. Consider physical space with only two dimensions, latitude and longitude. Now imagine that in a certain area a series of robberies occur. Here "robberies" is used as one example. Three useful questions for the police and the victims are the following.

1. Are the robberies the work of a single agent?
2. Is (are) the robber(s) from outside the area or do they live in that area?
3. If the robber(s) live in the area, where?

Let us suppose the following:

1. The robberies are work of one agent;
2. The robber lives in the general area of the robberies;
3. There are no statistics about previous robberies;
4. The exact time of each robbery is not known.

This means that the data set is composed of only the latitude and the longitude of each robbery,

$$Dataset = \{x_i, y_i\}_{i=1}^{N},$$

where x_i and y_i represent longitude and latitude of the ith robbery and N is the total number of the robberies. The problem at hand is, "Using only this information, is it

M. Buscema et al., *The Topological Weighted Centroid: A New Vision of Geographic Profiling*, Studies in Computational Intelligence 1095,
https://doi.org/10.1007/978-3-031-28901-9_6

possible to make a reasonable estimation about the place where the author might live? And is it possible to make the same estimation if the number of robberies is small? And if the collected data are about the places where an epidemic is spreading, is there a way that one might proceed to finding the source(s) (center) of the epidemic outbreak with the same mathematical algorithm?" Before presenting examples of several TWC applications, let us look at the various categories of problems already addressed by TWC.

6.1 Categories of TWC Applications

The applications of TWC to date are categorized as follows..

1. **Epidemics**
   (a) Dengue, Brazil 2001
   (b) Ebola, Africa 2014
   (c) COVID-19 Italy, USA, Brasil 2020, 2021, 2022
2. **Disease Dynamics**
   (a) Chikungunya, Italy 2007
   (b) West Nile Virus, USA 2008
   (c) Food Poisoning, Hawaii 2010
   (d) German HUS, Germany 2011
   (e) Listeria, USA 2011–2013
3. **Terrorism and Criminality**
   (a) Terrorist Attacks, Afghanistan 2009
   (b) Terrorism, Europe 2014–2016
   (c) Terrorism, North Africa 2014–2016
   (d) Terrorism, Middle East 2014–2016
   (e) Drug trafficking, London 2002
   (f) Florence Monster, 1968
   (g) Italian Unibomber, Italy 2006
   (h) Robberies and drug distribution in Denver, USA, 2010–2011
4. **Environmental and Political**
   (a) The Global Order from a geopolitical viewpoint, 1996
   (b) The Land of Fires, Naples, Italy 2013
5. **Cultural processes**
   (a) The Diffusion of Cultural Activities in Veneto, Italy, 2007
   (b) The Diffusion of Cultural Activities in Holland and Sweden, 2010

6. **Archaeology**
   (a) The Enigmatic Tablets
   (b) The Space of Megalithic Monuments
   (c) Tuscan Tribes.

6.2 Example Applications

Semeion Research Center, Rome, Italy, together with the Center for Mathematical and Computational Biology, University of Colorado, Denver, have applied TWC to many problems. What is presented below illustrate a partial range of the projects completed. The reader should consult the references for details.

6.2.1 Crime: Seven Robberies in Denver, Colorado, USA, December 2010

Between the end of December 2010 and January 2011, seven robberies occurred in Denver, Colorado. The police believed that the author of the robberies was the same agent and it was likely that the robber lived in the area. The data available were only the geographical position of the seven robberies. Thus, the algorithm had to work with 7 cases (latitude/longitude).

A TWC algorithm was applied to estimate the "hidden point" of the spatial distribution of events. Table 6.1 shows the data set and the TWC estimation and the real place where the offender was caught, which is listed as the last entry as TWC—Alpha. TWC identified a place (see Fig. 6.1), the "hot" (red) area, which the algorithm

Table 6.1 Robberies

Addresses	Long	Lat	Type of data
138 N. Del Mar Cir.	-104.844635	39.728157	place of the robbery
3106 S. Parker Rd.	-104.854683	39.660419	place of the robbery
3122 S. Parker Rd.	-104.851326	39.660076	place of the robbery
700 N. Dayton St.	-104.874985	39.726479	place of the robbery
1878 S. Havana St.	-104.865799	39.682446	place of the robbery
2630 S. Parker Rd.	-104.862648	39.668732	place of the robbery
669 North Peoria Street	-104.849533	39.727028	place of the robbery
TWC Alpha	-104.865799	39.682446	TWC estimation of robber house

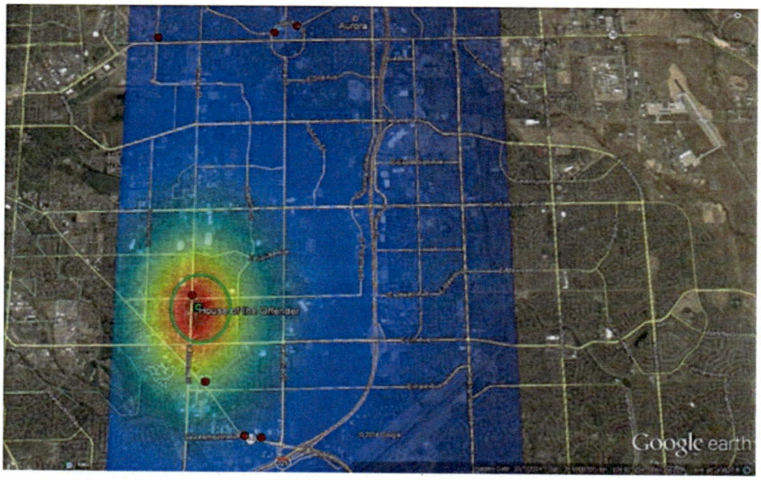

Fig. 6.1 Robberies—Denver, Colorado

suggested to investigate as a possible residence of the robber. On the same map, it was later found, the actual place of residence of the robber. In this case, the "hidden area", estimated by the TWC algorithm, was within a radius of about 100 m from the residence of the robber as was later discovered.

6.2.2 Crime—Italian Unibomber, Veneto, Italy, 1991–2006

In 2006 the Italian Unibomber was still unknown after 15 years of research and 33 attacks with homemade bombs against children. Many gravitational algorithms beyond traditional techniques were applied by Italian police without any success. TWC was applied using only the positions of the 33 attacks. 800 m from the estimated TWC spot was identified a shed in which were found a pair of scissors and many special chemicals used for many of the attacks. After the seizure of the shed made by the judiciary, there were no further attacks (see Figs. 6.2 and 6.3).

The traditional police methods lead to searches at the locations given in black. The green is another profiling algorithm location. The TWC location was 800 m from the spot where the unibomber used to put the bombs together.

6.2.3 Crime—24 Drug Arrests—Denver, CO, USA, 2010–11

Twenty-four arrests were made for drug distribution from September 2010 until April 2011 were given to be analyzed that occurred in Denver, Colorado. The question was whether TWC could locate the dealer who distributed the drugs to those who were

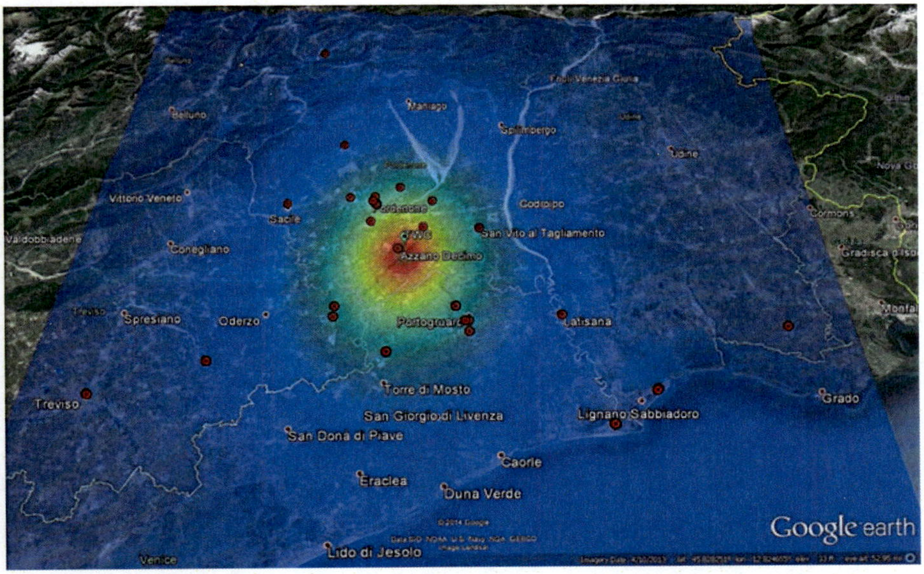

Fig. 6.2 Italian Unibomber

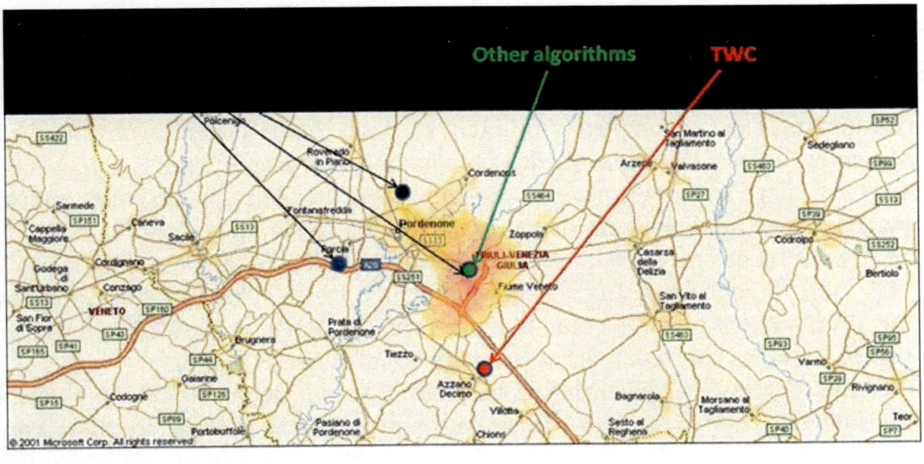

Fig. 6.3 Unibomber predicted location

arrested. TWC was applied and the algorithm identified a location that was less than one mile from the house where the dealer was caught (see Fig. 6.4). The drug distribution arrests are in red.

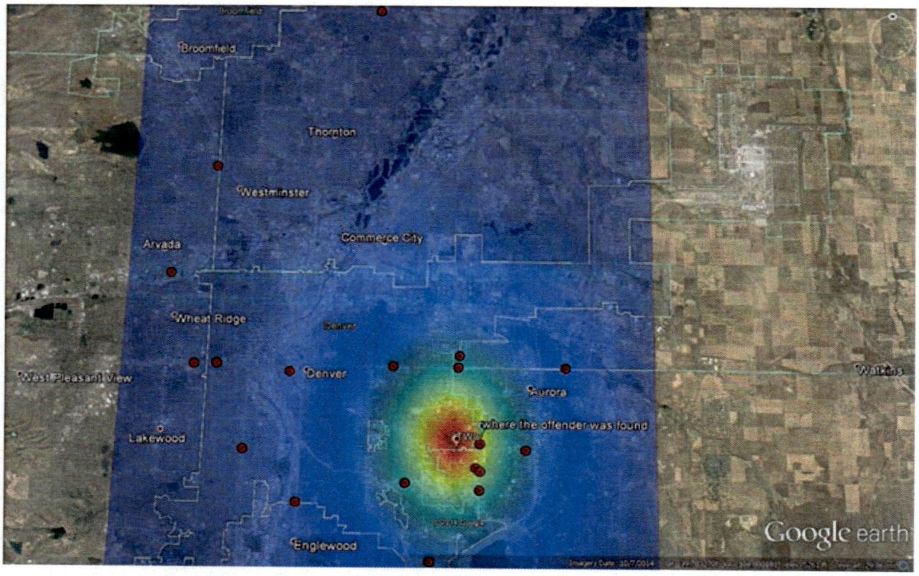

Fig. 6.4 Twenty-four drug distribution arrests—Denver, Colorado

6.2.4 *Cocaine Trafficking in London, 2006: A Real Case*

We now consider a completely different problem, based on real world data and presenting a typical example of what is commonly meant as a 'hard' analytical task. We start from a map of the places in London where the Metropolitan Police arrested persons caught trafficking in cocaine in a four months period [1] (see Fig. 6.5).

Looking at this map, it is pretty clear how the distribution of the entities is hardly believable to be random. In the following picture we can observe how, computing the TWC—Alpha, it can be seen to move from the center of mass down towards a group of entities situated in the borough of Bromley (see Fig. 6.6).

According to other independent research [1], conducted using different data, it turns out that Bromley and the City of Westminster appear as two of the most representative boroughs in London for cocaine trafficking. If we generate the scalar field of this distribution, using TWC, we have an optimal representation of the proximity to cocaine trafficking in London, Bromley and the City of Westminster are the two foci of cocaine distribution (see Fig. 6.7).

Now, projecting the same map according to the gradient, we are also able to distinguish the different clusters of cocaine trafficking in London (see Fig. 6.8).

An interesting analysis was done next to obtain a hypothesized conjecture of the most probable trajectories connecting all such drug dealing places (entities) together. This analysis was done by building the nonlinear MST, based on the paths connecting the entities, (see Fig. 6.9).

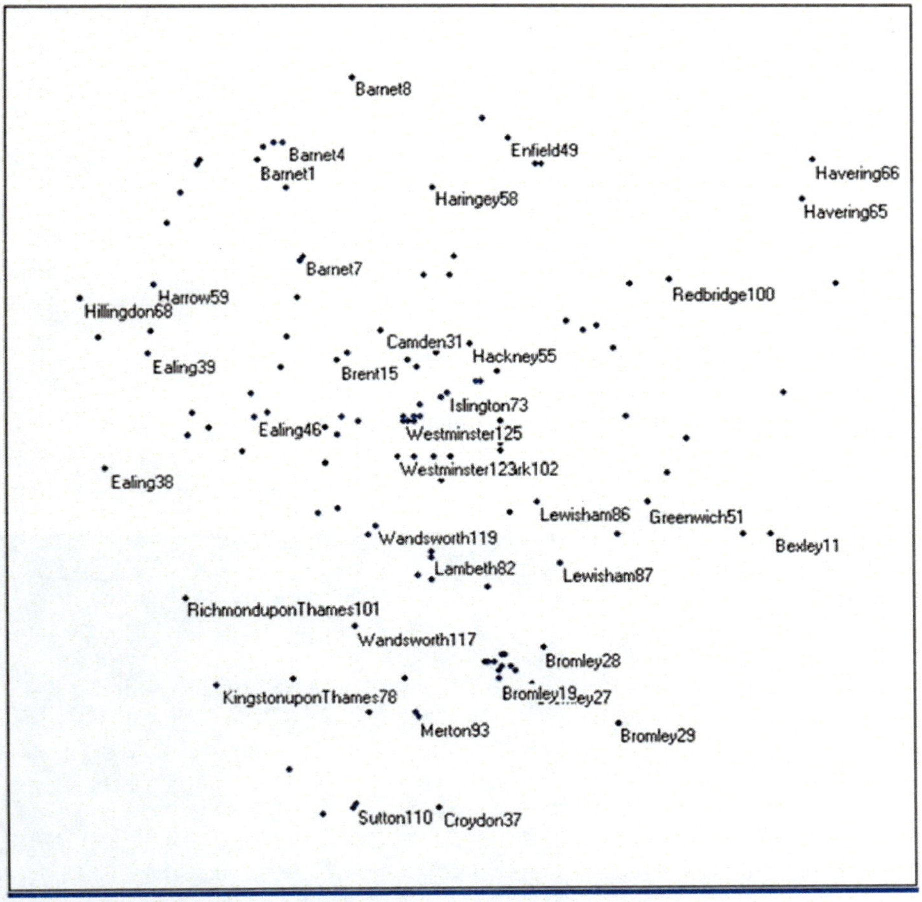

Fig. 6.5 Arrests—London Cocaine trafficking over 4 months

This graphical map could, at first sight, look too complex to be readable in any sensible way. To help understand the connections, an alternative representation is provided by TWC, the implementation of the Maximally Regular Graph (MRG) algorithm. The MST representation displays a subset of the links that are necessary to capture the fundamental connections between the observed entities [1–5] (see Fig. 6.10).

It is possible, by looking at MRG graph, to hypothesize that the cocaine delivery network in London starts from the borough of Bromley (TWC above and to the right of the graph), and has its wholesale hub in the borough of Merton and finally arrives, through Lambeth, at the big, final customer market of the City of Westminster and subsequently to the other boroughs of London (relatively minor final customer markets).

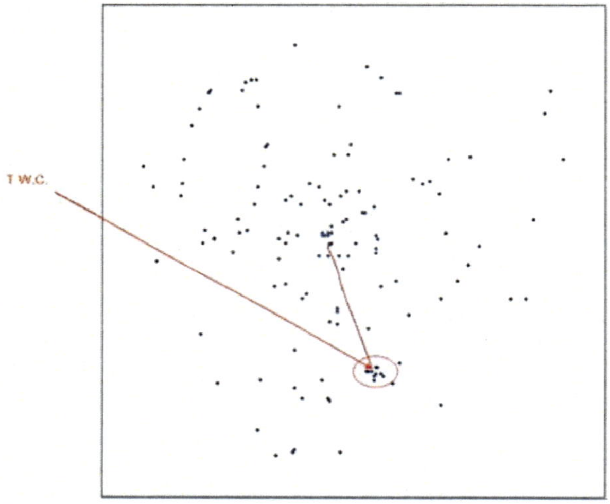

Fig. 6.6 Center of mass versus TWC-Alpha point

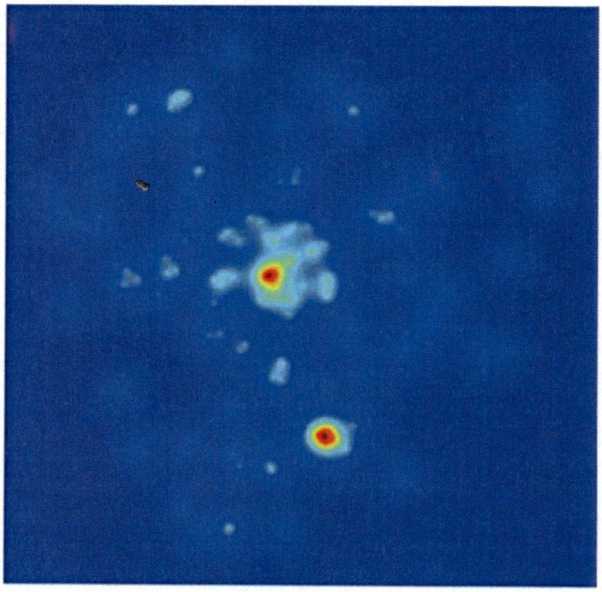

Fig. 6.7 Scalar field—London Cocaine arrests

This example explains pretty clearly what we mean by a spatial relational analysis, and to what extent our approach exploits a particular form of space semantics. The task of reconstructing the spatial articulation of cocaine trafficking in a large metropolitan area in London by looking only at events that locate the capture of drug dealers in a given time period looks somewhat hopeless at first sight. Nevertheless, it is clear that, insofar as a real drug trafficking network is operating across the London territory, it must have an inherent spatial organization that has to emerge to some degree from the observable data. The question becomes whether a small scattering

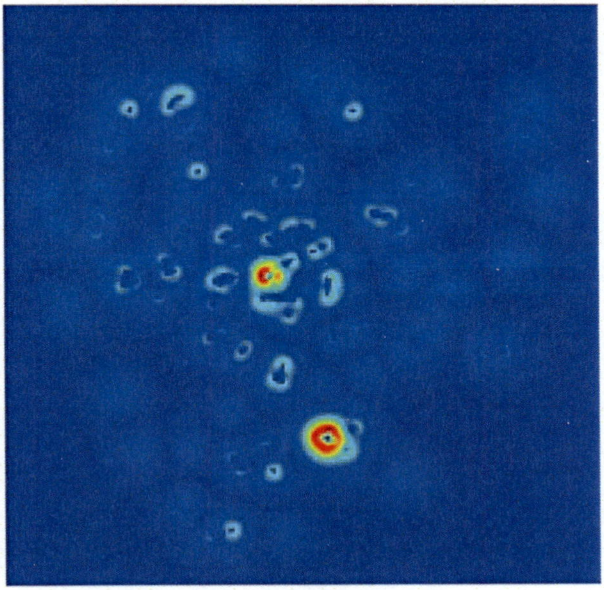

Fig. 6.8 Gradient map—London Cocaine arrests

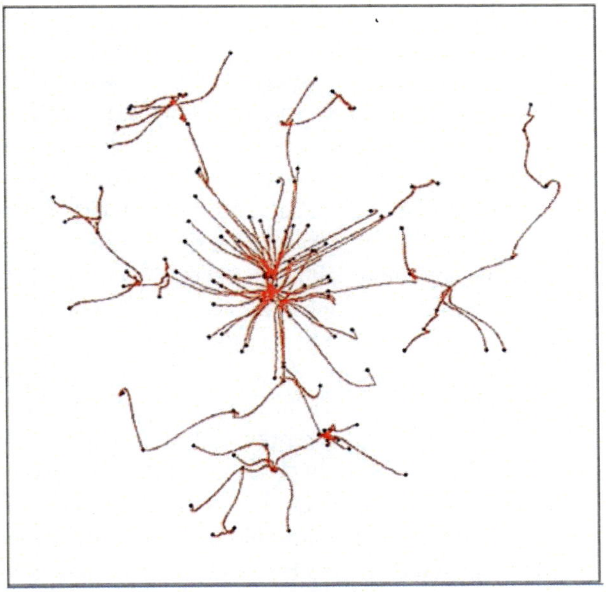

Fig. 6.9 Nonlinear MST—London Cocaine arrests

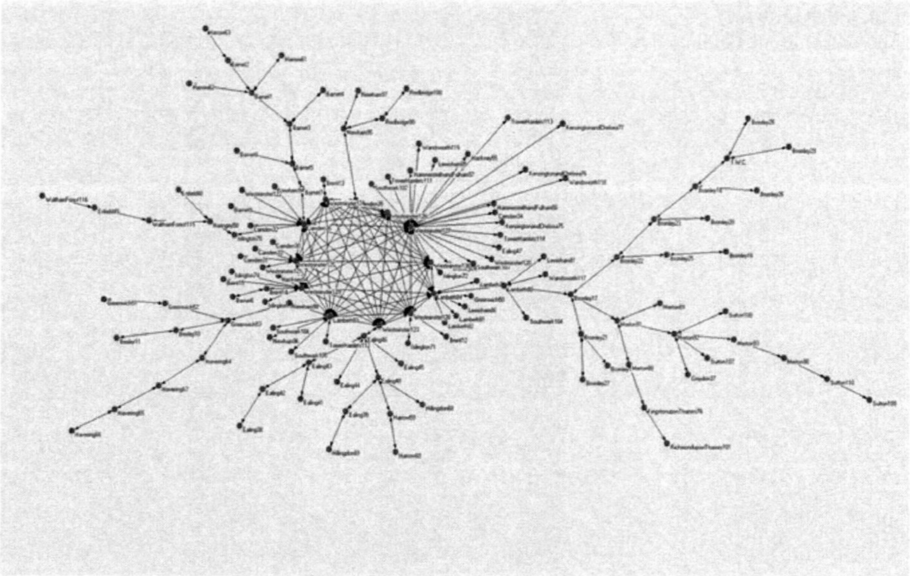

Fig. 6.10 Maximally regular graph—London Cocaine arrests

of points, like the data made available for this example analysis, might be enough to reconstruct the basic structural properties of the drug dealing network. If we consider the available data (the entities) as isolated points in space, that is, as random draws from a given, underlying distribution, the task tends to be very difficult if not hopeless. But if, on the contrary, such data are thought as context-sensitive, that is, a given event (the arrest of a drug dealer) produces complex spatial effects on the entire geography and organization of drug trafficking (like a propagation wave following a shock), then it is clear that the observed spatial distribution of entities carries a meaning, and in fact governed by underlying semantics that can be reconstructed on the basis of a few well chosen reference cases. From this perspective, it should also be clear in what sense the various entities "negotiate" their position in the underlying, warped representational space. Some of the dealers are more tightly connected to some of the others, and thus the event of the capture of a given dealer produces asymmetrical effects in the whole network, and affects differently, the as of yet uncaught dealers in different ways, thereby causing complex adjustments that will reflect in the distribution of the next set of drug arrests. This is the kind of hidden, structural information that we need to learn to exploit to be able to reconstruct the true semantics of the phenomenon under study.

Remark 6.1 Our analysis shows how relatively elementary mathematics may help decision makers in situations characterized by a substantially limited amount of information, and moreover how the mathematics of complex systems can improve the level of accuracy in the analysis of spatial phenomena, compared with the level that can be achieved with a classical statistical toolbox. We have illustrated a few applications that deliver interesting and encouraging results, but we are at the same time aware that the methodology presented here needs more, careful verification and validation on further logically consistent and empirically relevant problems. We are convinced that this example is but one example of the actual range of problems from many different disciplines that may be usefully addressed by means of the TWC methodology. This is seen in the subsequent examples that are presented.

6.2.5 *Epidemics—Ebola Virus Outbreak of 2014*

March of 2014 was the beginning of an Ebola outbreak that broke out in Guinea. August 30, 2014, the known cases in West Africa (Guinea, Liberia, Sierra Leone, Nigeria, Senegal and the Democratic Republic of Congo) were over 3,000 with a mortality rate of over 50%. The site

$$http://healthmap.org/ebola/\#$$

reported, for each day, the situation. However, there were no places on the internet nor were there reports about the geographical coordinates of each case of Ebola. Since TWC requires the spatial location (latitude/longitude) of events such as these, this Ebola analysis took the descriptions of where the Ebola cases occurred and approximated the geographical coordinates of 43 places where at least one case was reported to have occurred (see Fig. 6.11). These data are inaccurate and incomplete, but the goal was to:

1. Estimate the location of the center of the main outbreak of this new epidemic;
2. Estimate where and how it will spread.

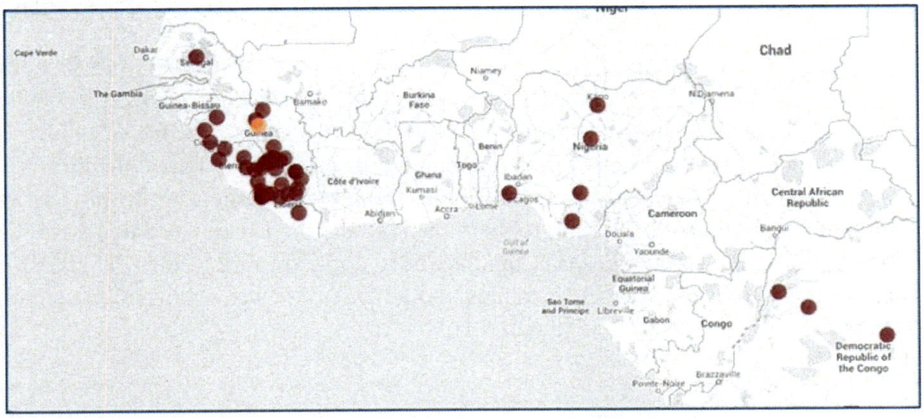

Fig. 6.11 Map of the Ebola cases

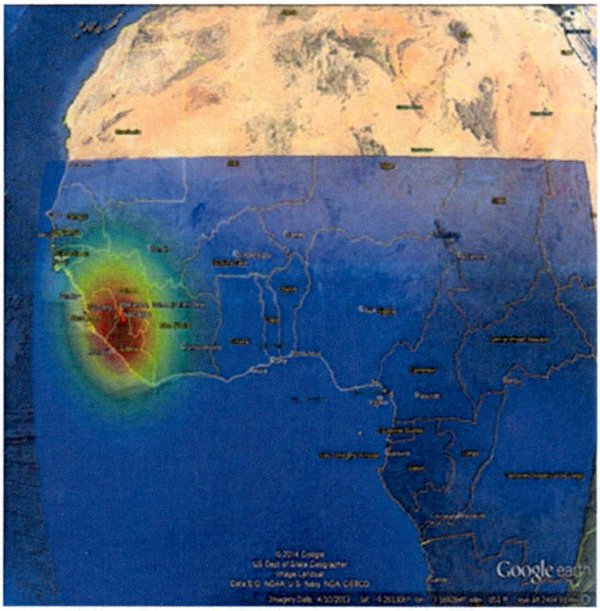

Fig. 6.12 Ebola hot spots

Table 6.2 Ebola cases from September to October 2014

Ebola cases	1st september 214	6th october 214	Increasing (%)
Guinea	648	1199	85.03
Sierra Leone	1026	2437	137.52
Liberia	1378	3834	178.23
Nigeria	17	20	17.65
DR congo	24	70	191.67

The TWC processing of the 43 positions identifies two possible outbreak sources in Guinea. The first TWC outbreak "hot spot" is located 30 km from the place where it was thought to be the real "zero case" (see Figs. 6.11, 6.12). Table 6.2 shows the increasing number of cases of the epidemic 3 months after the data collection.

The algorithm that was applied to Ebola (see [6]) is the same algorithm that was applied to the Denver, Colorado robbery data set and the London cocaine arrest data set. And it is the same algorithm that has been applied to more than 10 epidemics and to more than 10 data set of crimes and terrorist attacks, published in scientific literature (Figs. 6.13, 6.14, 6.15 and 6.16).

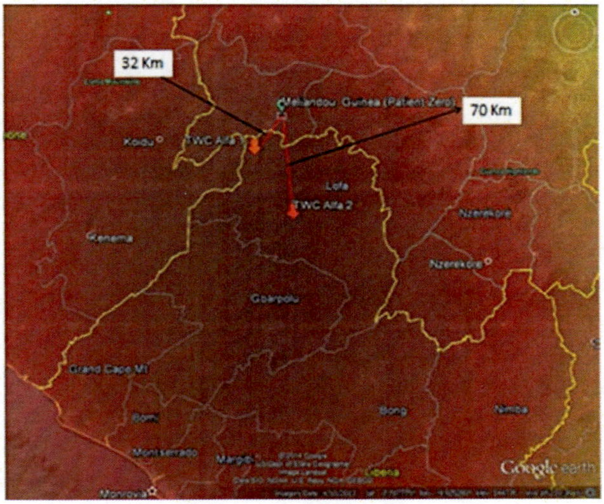

Fig. 6.13 Close-up of the hot area

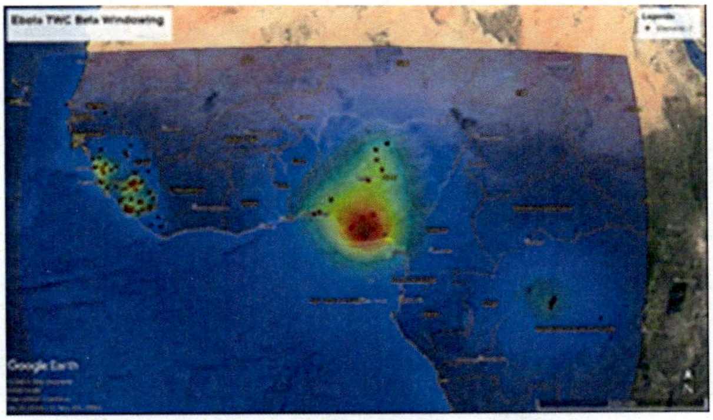

Fig. 6.14 Hot areas of TWC Beta

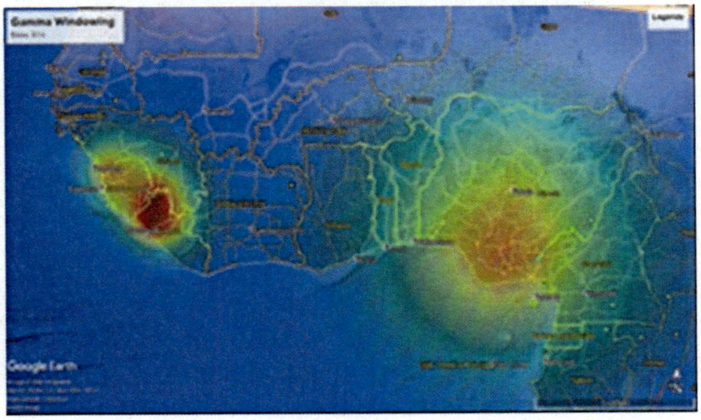

Fig. 6.15 Hot spots of TWC Gamma

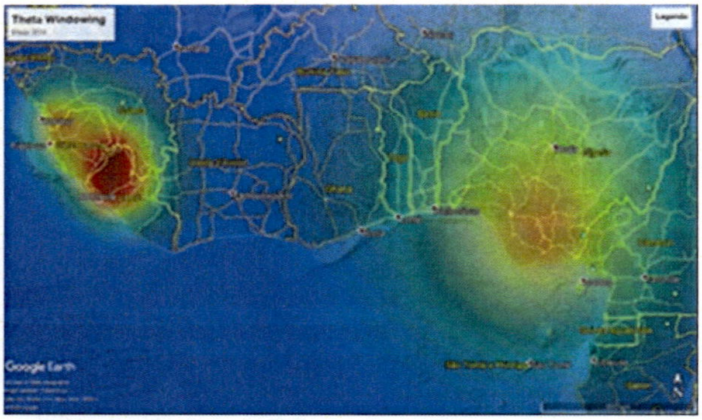

Fig. 6.16 Hot spots of TWC Theta

6.2.6 Epidemics—Dengue Fever in Brazil 2001

There was a severe Dengue fever outbreak in the state of São Paulo, Brazil beginning in 2001 that lasted for approximately 5 years. Thus the data set was robust. Moreover, the data reported the coordinates of 54 towns infected by Dengue fever. The distance between the real outbreak and the estimated one by TWC was about 4.5 km (see Fig. 6.17).

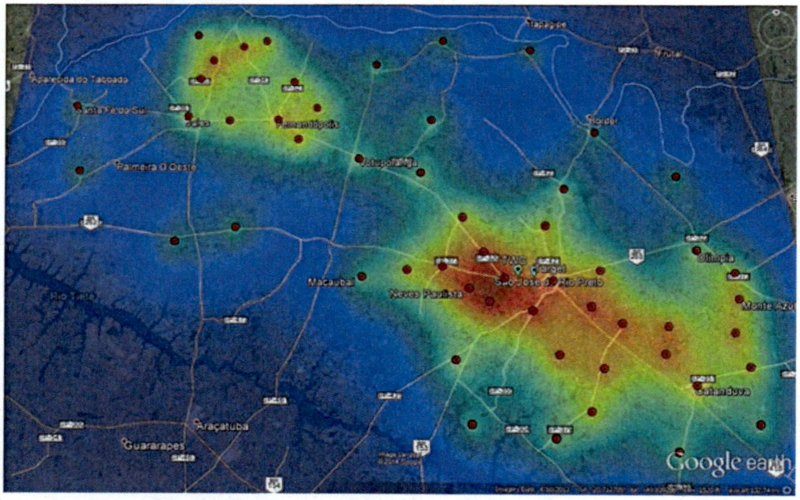

Fig. 6.17 Dengue outbreak 2001 in the state of São Paulo, Brazil

6.2.7 *Epidemics: COVID-19 USA—2021 Analysis*

TWC was used to analyze the COVID-19 outbreak in the United States of America. In this analysis the COVID-19 cases in USA on March 26, 2020 as provided by Johns Hopkins University was used. The COVID-19 outbreak is mapped by the TWC method. The prediction obtained by TWC was able to capture some features of the pandemic spread using this early data. Although TWC used the geographical distance from the latitude and longitude coordinates, the results indicate that one of the main paths of diseases spread is arguably airline routes. For this analysis a large set of data was available.

6.2.7.1 TWC-Windowing

The capturing of local and global effects of the COVID-19 epidemic was used given the vast expanse of the geographic area (see [7]). The TWC analysis in this paper used $N = 3,146$, $K = 50$, $\frac{N}{K} \approx P = 62$. The analysis constructed a graph the relationship between the cases which, visually, appeared to follow airline routes. This was further analyzed below.

6.2.7.2 TWC-Alpha Map

Figure 6.18 shows the results of the TWC-Alpha algorithm. The hotspots (the origin of the outbreak) are shown in this figure. Two possible outbreak origins are located in west, the states Washington and California. The northwest hotspot that originated in Washington state is expanded toward the east and covers the neighboring states. There are several isolated hotspots, for example there is one hotspot in northeast.

Three clusters were found, labeled A, B and C (see Fig. 6.19), which can be considered as attractors according to the Markov Chain inferred from TWC-Theta. The nonlinear MST was applied. These points correspond to the attractors of the dynamics of the whole process. They are obtained by transforming the nonlinear distance matrix (from which the nonlinear MST is derived) into a discrete time Markov Chain. Figure 6.19 shows the attractors points for clusters A, B and C on the map of the confirmed cases. Figure 6.20 shows the attractor and airports (also see [7]). Given the strong correlation between the attractors and the airports, the hypothesis is that airplane routes are important sources of the spread of covid.

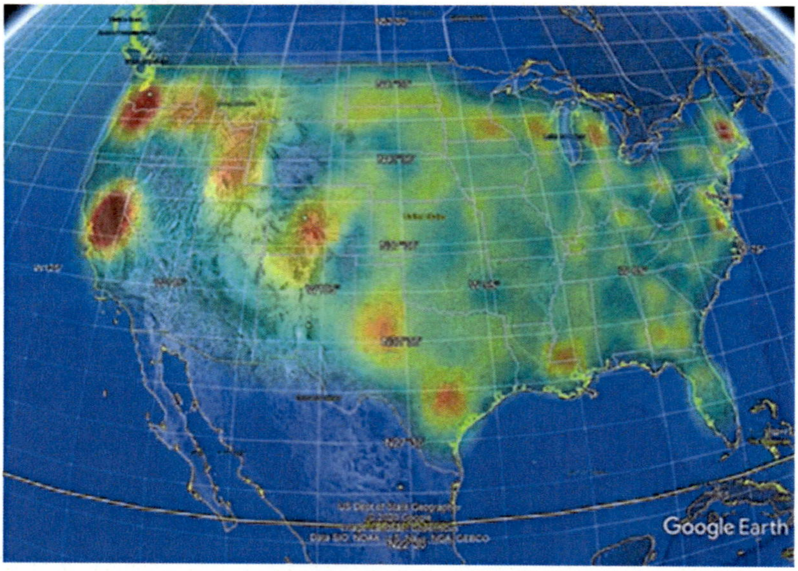

Fig. 6.18 The result of TWC-Alpha: COVID-19 USA March 26, 2020

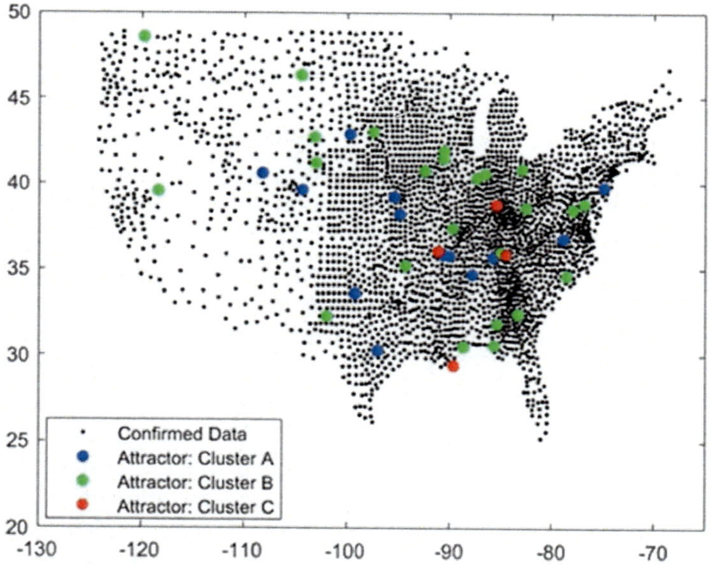

Fig. 6.19 Attractors on USA COVID-19

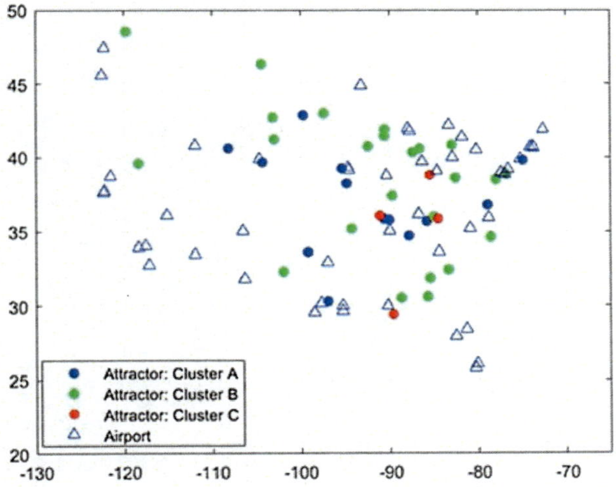

Fig. 6.20 Attractors and airports

6.2.8 *Disease Dynamics—The German HUS in May 2011*

The German HUS (Escherichia Coli mutation) occurred in Germany on May 2011. At the end of May only 13 German cities registered that they had infections with at least one case. German authorities identified the outbreak in Hamburg (see [8]), but on June 18 they found the evidence that another source of the outbreak was in Frankfurt. TWC algorithm was applied on May 28, 2011 at University of Colorado Denver. TWC has estimated, using only the 13 reported cities, the sources of the outbreak in Hannover and Frankfurt, twenty days before of the German authorities and the predictive estimations were validated three months after (see Figs. 6.21 and 6.22).

The first 13 cities infected as reported May 2011, are the red points. There are two TWC $PMPs$, one south of Hamburg and one at Frankfurt. TWC—Beta was implemented May 2011 on the data for the 13 infected cities. TWC—Gamma pre-

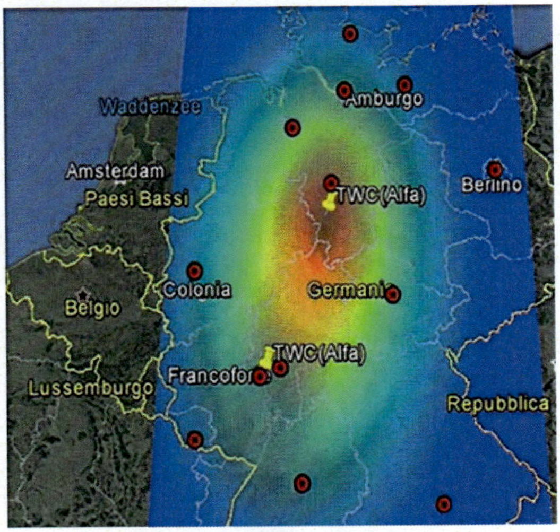

Fig. 6.21 German E-Coli—TWC Alpha May–June 2011

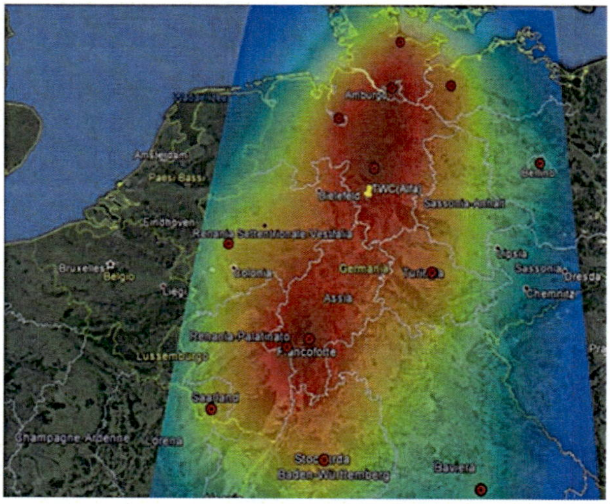

Fig. 6.22 TWC Beta

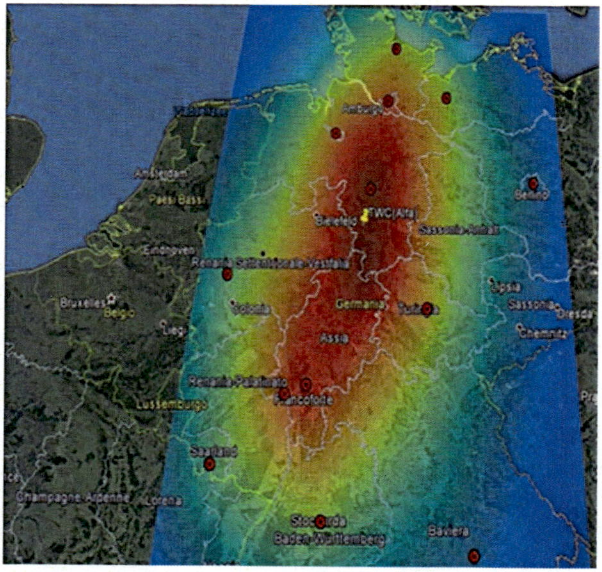

Fig. 6.23 TWC Gamma

diction of the distribution of the outbreak as also done May 2011. The TWC May 2011 prediction of the medium term spread of the E. Coli outbreak are found in the figures for TWC—Gamma and TWC—Theta. The long term spread of the outbreak can be found in the figure depicting TWC-Iota. Figure 6.22 has the distribution on the left done May 2011, which can be compared with the actual distribution map as produced by the German Health Department depicted on the right (Figs. 6.23, 6.24, 6.25, 6.26, 6.27 and 6.28).

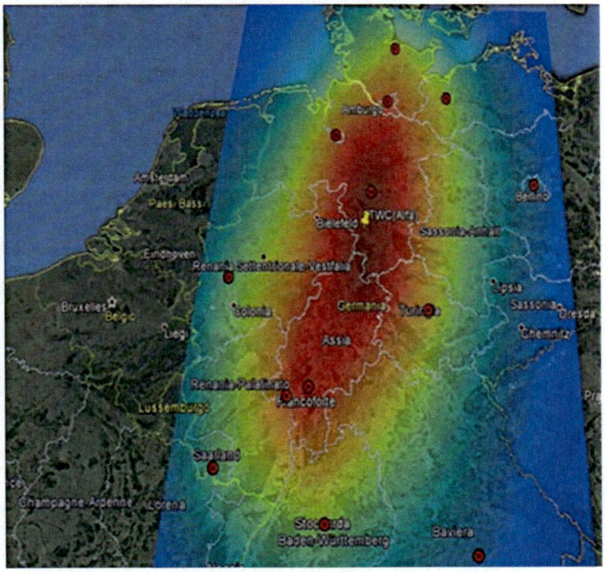

Fig. 6.24 TWC—Gamma

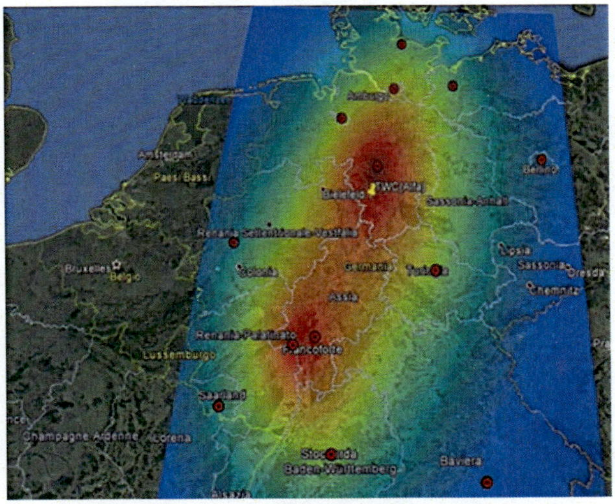

Fig. 6.25 TWC—Theta

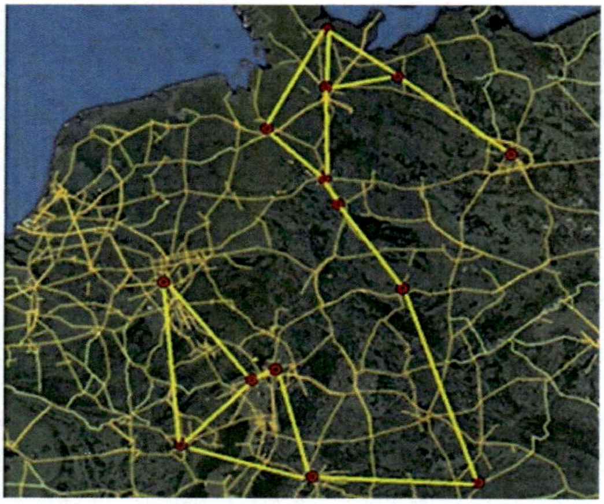

Fig. 6.26 German E. Coli outbreak—TWC Theta

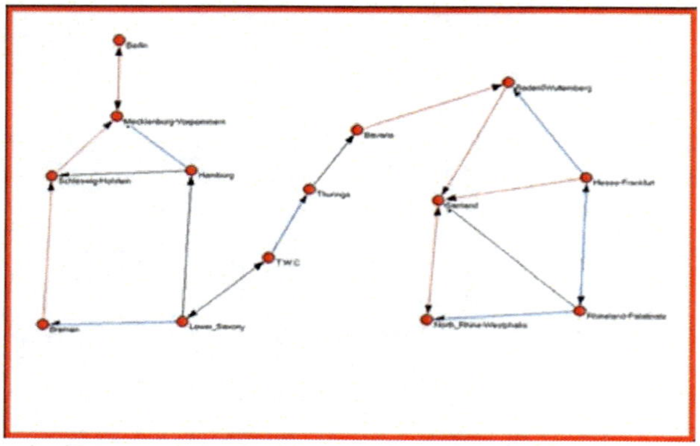

Fig. 6.27 German E. Coli outbreak—Directed graph

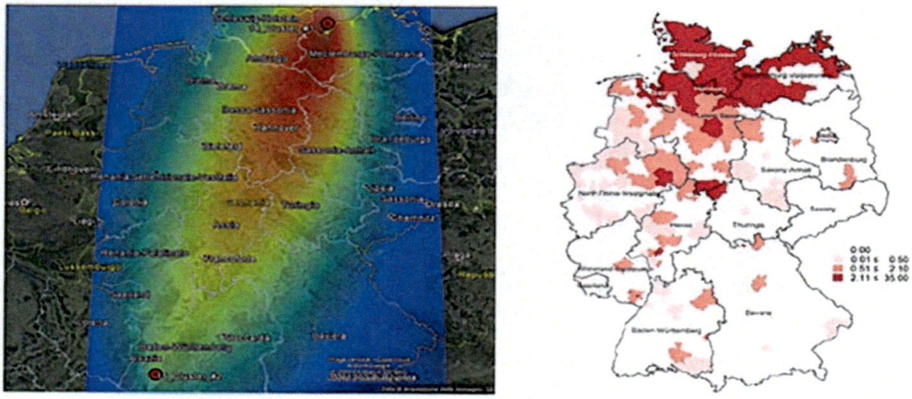

Fig. 6.28 German E. Coli TWC—Iota comparison

6.2.9 Disease Dynamics: Listeria USA 2011

The listeria epidemic that was linked to cantaloupes occurred in United States in 2011 and was analyzed retrospectively using TWC. The first 72 calls to the poison centers around the United States dealing with questions about cantaloupes from September 1 to December 31 collected from the National (USA) Poison Data System and TWC was applied to estimate the first outbreak of this epidemic. It is noted that the 72 poison center cantaloupe calls were used since the Colorado Department of Health did not make the latitude/longitude data of the actual cases available. The National Poison Data System data was a *national* data set rather than a localized Colorado data set. Even though the data set was at a national level data, the TWC estimation resulted in the identification of a probable area 50 km from the real outbreak, a cantaloupe farm close to Holly, Colorado, (see Fig. 6.29). Moreover, the TWC maps were run day by day to see if Holly could be identified within 50 km within a time frame less than the time it took the Colorado Department of Health to identify Holly as the location. TWC was able to locate an area with a 50 km radius using 16 days of data. The Colorado health department took much longer to identify the farm from which the listeria came. While TWC identifies a probable area rather than a specific location, this analysis shows that TWC could be used to focus the search especially in the initial stages of an outbreak such as listeria (see [9]).

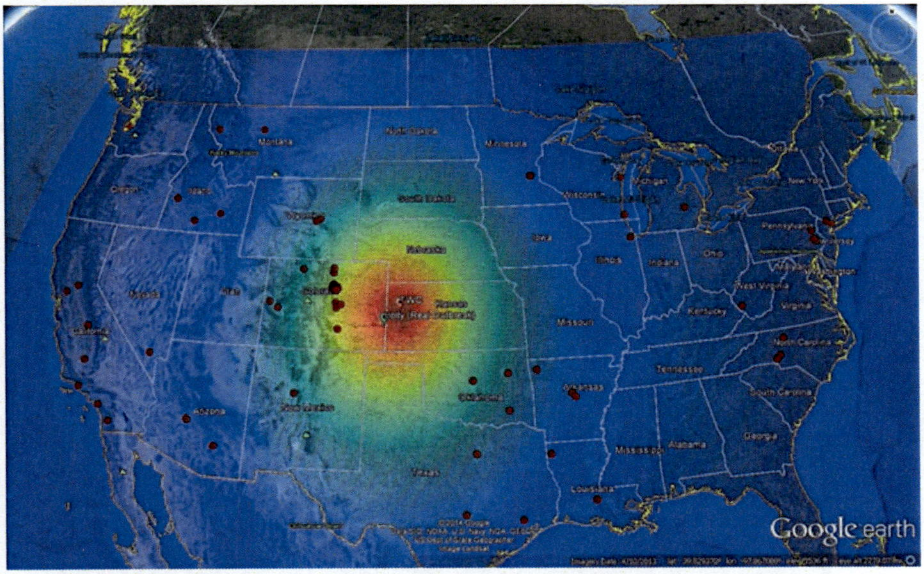

Fig. 6.29 Listeria outbreak

6.2.10 Disease Dynamics: Hawaii, 2010 Food Poisoning

Oahu, Hawaii, experienced a food poisoning outbreak in 2010. The data that was given to be analyzed consisted of the location of 1,245 cases collected systematically over 12 months. The data set was collected and the request given to the poison center that serviced Hawaii, which happened to be Rocky Mountain Poison Center at that time. Figure 6.30 shows the geographical distribution of cases of the Oahu food poisoning epidemic at the end of 2010 and Eq. 6.1 shows the distribution of the new cases each month of 2010.

Month	*NumberofCases*
January	108
February	109
March	114
April	98
May	79
June	79
July	93
August	109
September	92
October	134
November	114
December	108

(6.1)

Table: Distribution of the Hawaii food epidemic in 2010 in each month.

Figure 6.31 shows the cases of January 2010 with the two TWC—Iota meta-clusters points found. TWC—Iota global field detected these clusters analyzing only the January data points. From Fig. 6.32, it can be seen that the two meta-cluster points

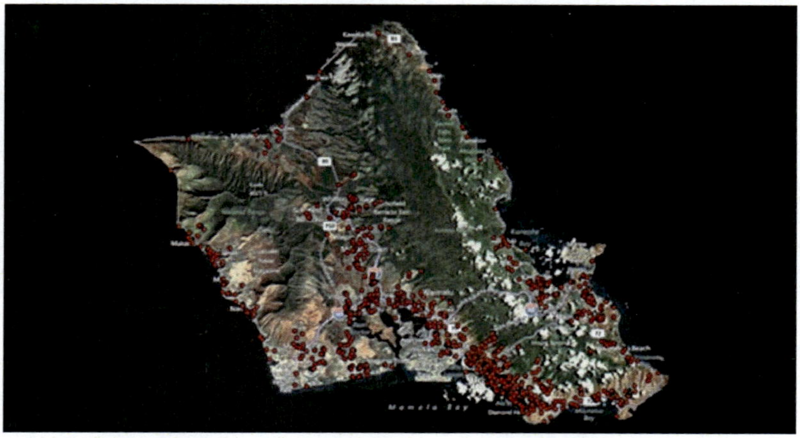

Fig. 6.30 Hawaii food poisoning locations

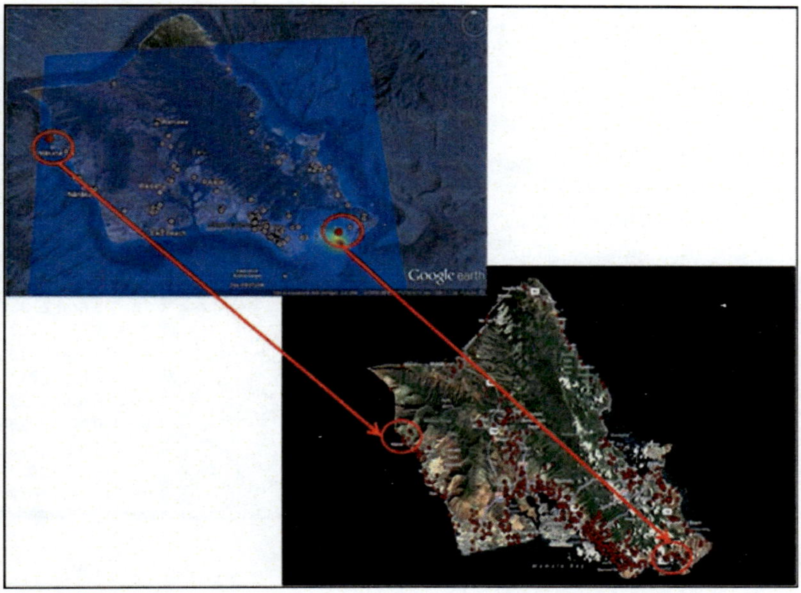

Fig. 6.31 TWC—Iota meta-clusters (January 2010)

Fig. 6.32 Total cases of 2010

are located in areas with no cases or few cases. If we compare this map with the map reporting all the cases that occurred up until the end of 2010, the algorithm predicted the two meta-clusters points with a high concentration of new cases (see Figs. 6.31 and 6.32). It is noted that the two circled areas (in red) were predicted using only the January of 2010 data. This application is an example of the capability of TWC—Iota meta-clusters to detect interesting points of the development of a given process some of which are outside the given area of the data points.

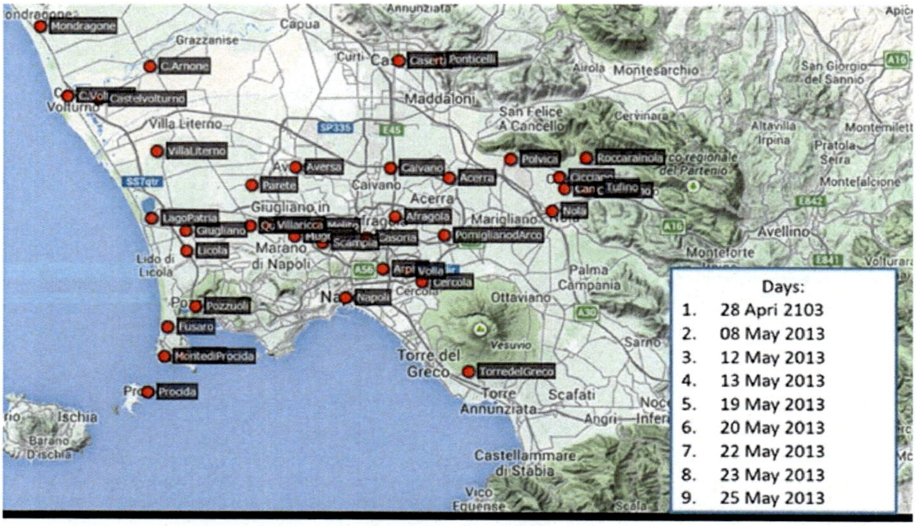

Fig. 6.33 Fires over 9 days

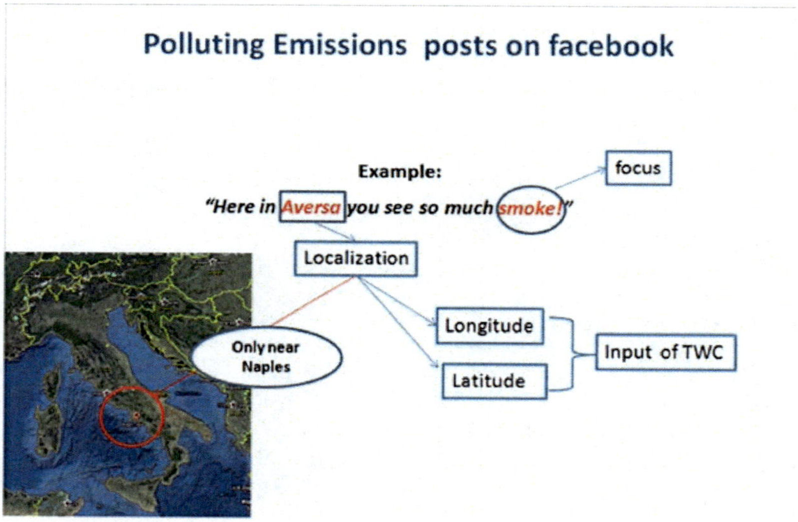

Fig. 6.34 Facebook postings

6.2.11 Environment—Fire, Italy, 2013

Two Italian provinces, Naples and Caserta, at beginning of 2013, experienced the phenomenon of toxic burning, that became an emergency. There were over 6,500 fires in less than two years that occurred in the municipalities of Caserta and Naples. Meanwhile there was a growing movement of citizens who were alarmed and worried about the increase in the number of cancer cases in polluted areas (see Fig. 6.33).

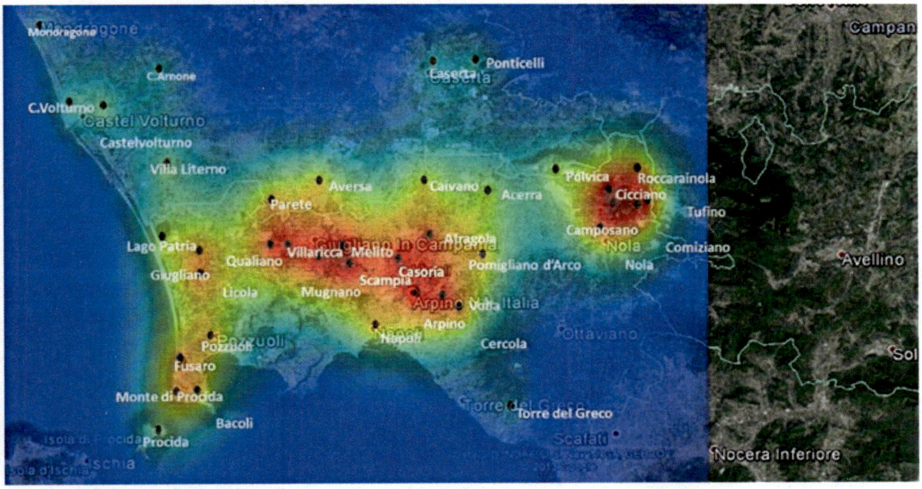

Fig. 6.35 TWC results

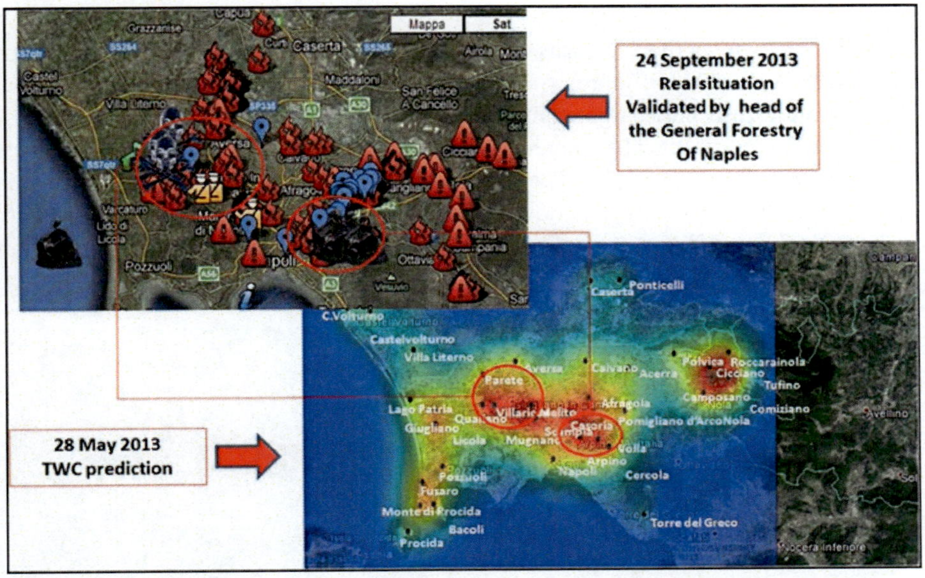

Fig. 6.36 TWC results versus forest services

There were 39 posts warning of fires from Facebook in nine days, from 28 April 2013 to 25 May 2013). . For each warning and post, the coordinates of fires were identified (see Fig. 6.34).

The TWC algorithms were implemented to try to help with the location of fire fighting resources to combat the burning (Fig. 6.35). On the 24th September 2014, the head of the General Forestry of Naples found a real situation completely similar to the map predicted by TWC, three months before (Fig. 6.36). On the 31st of October,

2013, two butchers were arrested at Aversa and 4 companies and 15 citizens were fined because of the toxic burning exactly in the places predicted by TWC algorithm.

6.3 A Brief Comparison Between the Geometric and Topological Approaches

Three TWC applications were selected for this brief comparison between TWC and Geographic Profiling: the German HUS (May 2011 see [8, 10]), the Oahu, Hawaii food poisoning outbreak of 2010, and the Ebola epidemic (Africa 2014–2018 see [6]) . We show here only a qualitative comparison, but a detailed quantitative analysis can be found in the already quoted papers.

6.3.1 German HUS (May 2011)

Figure 6.37 shows the 13 towns where epidemic events of German HUS occurred up to 28 May 2011. Figures 6.38, 6.39, and 6.40 show, respectively, the scalar fields found by Canter algorithm, and TWC-Beta. It is evident what these different approaches detect. The Canter and Rossmo algorithms focus the two big clusters where the epidemic is active. The geometrical closeness of the 13 towns is the main criteria of this clustering. Hannover and Hamburg seem to be the center of the first cluster and Frankfurt seems to be the center of the second cluster. TWC-Beta shows,

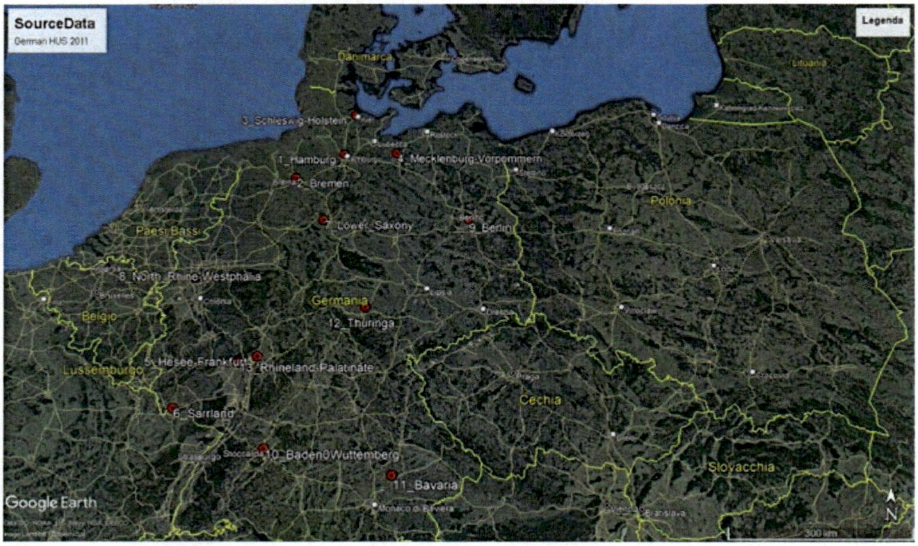

Fig. 6.37 German HUS epidemic

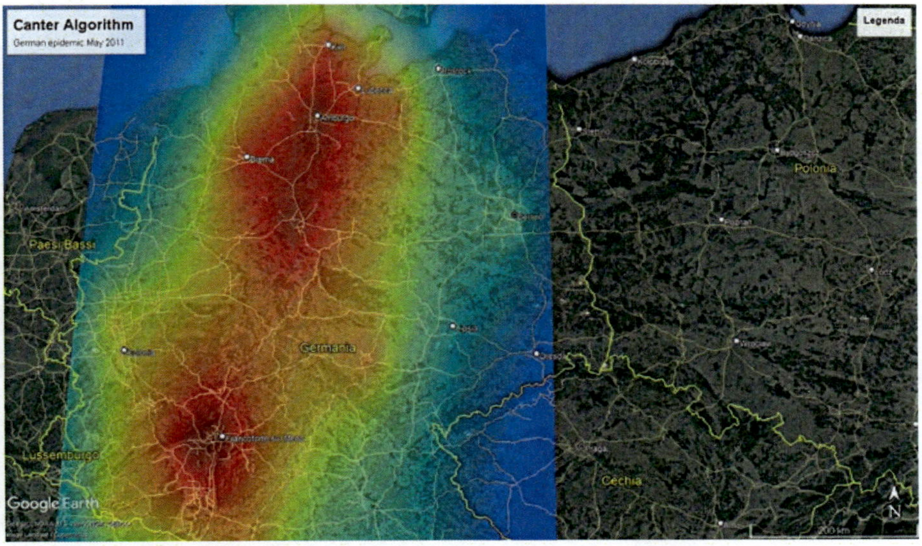

Fig. 6.38 German HUS epidemic, Canter scalar field

more or less, the same behavior of the previous ones. TWC-Alpha, instead, focuses, with a high accuracy, only a specific area under Hannover. From the point of view of TWC-Alpha, the outbreak of the epidemic was located in south area, between Hannover and Kassel, and only in a second time it moved toward Hannover and Hamburg from one side and toward Frankfurt from another side, as TWC-Beta, Canter and Rossmo algorithms hypnotize.

The following figures show the scalar fields rebuild by the four algorithms in a 3−dimensional surface. Canter and TWC-Beta 3−dimensional maps are very similar, Rossmo map is also similar to the previous one, but not smooth, while TWC-Alpha is completely different from the others. The first 13 towns is shown as red dots where at least one event of the epidemic occurs up to May 28, 2011 (Figs. 6.41, 6.42, 6.43 and 6.44).

6.3.2 Oahu Food Poisoning Outbreak 2010

Oahu food poisoning outbreak occurred during the entire year of 2010 with 1,244 events that were collected. The first figure shows the geographical distribution of the data. The next two figures show the scalar fields found by Canter and Rossmo algorithms. The next two figures show, instead, the scalar fields according to TWC-Alpha and TWC-Beta (Figs. 6.45, 6.46, 6.47, 6.48 and 6.49).

This application also shows that Canter and TWC-Beta algorithms have a similar behavior, while the Rossmo algorithm appears to be a very generic map. In any case,

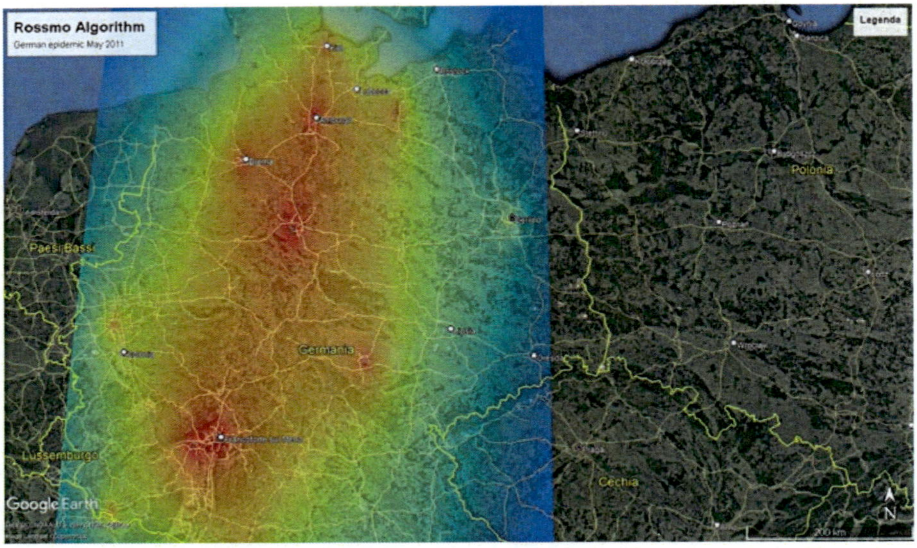

Fig. 6.39 German HUS epidemic, Rossmo scalar field

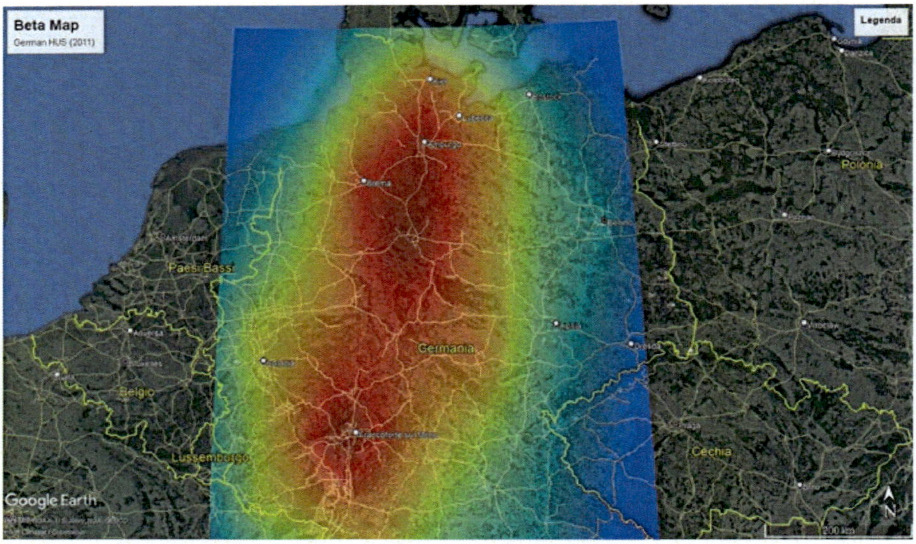

Fig. 6.40 German HUS Epidemic, TWC-Beta

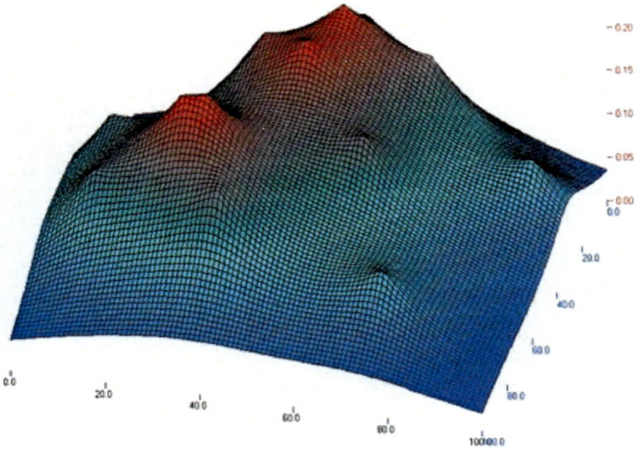

Fig. 6.41 Canter algorithm

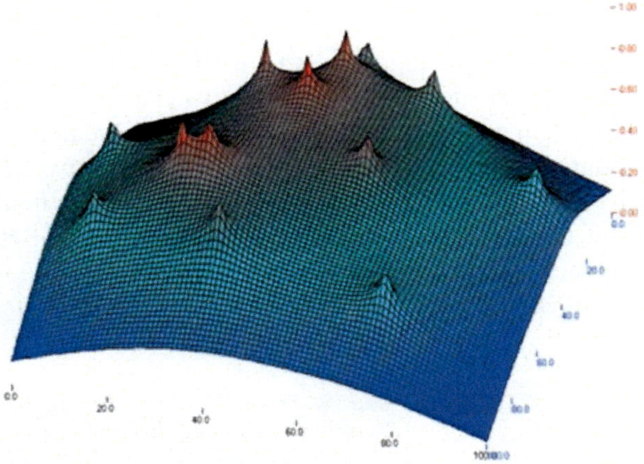

Fig. 6.42 Rossmo algorithm

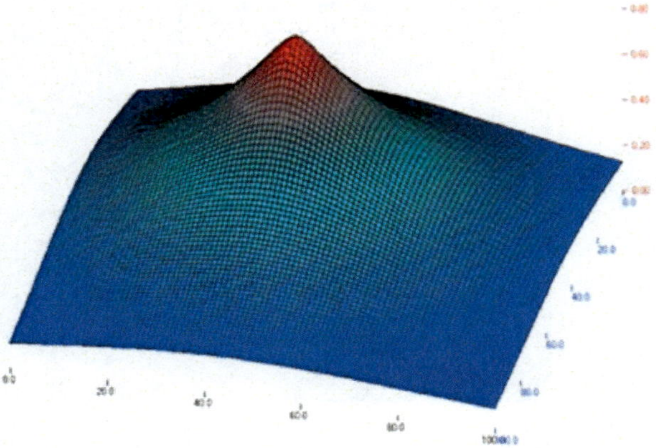

Fig. 6.43 TWC-Alpha

Fig. 6.44 TWC-Beta

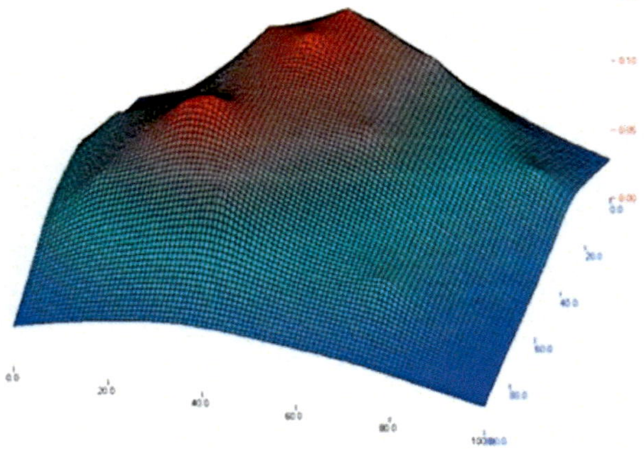

all these three algorithms focus on the area where the frequency of the events is higher. However, TWC-Alpha, instead, shows a different behavior. Its scalar field detect a specific area of high risk just to the north of the other algorithms' area and outside of the zone with high frequency of cases (Figs. 6.50, 6.51, 6.52 and 6.53).

The following four figures show the scalar fields rebuild by the four algorithms in a 3–dimensional surface. Canter and TWC-Beta 3–dimensional maps are very similar, Rossmo map is also similar to the previous one, but with less specificity, while the shape of TWC-Alpha is completely different from the others.

Fig. 6.45 Oahu Food poisoning data (red dots) Jan–Dec 2010

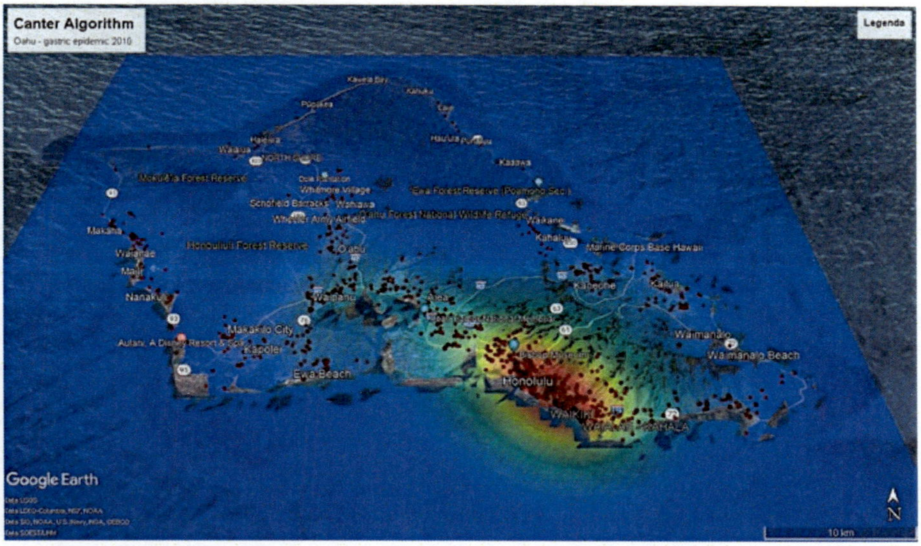

Fig. 6.46 Canter algorithm, Oahu gastro epidemic Jan–Dec 2010

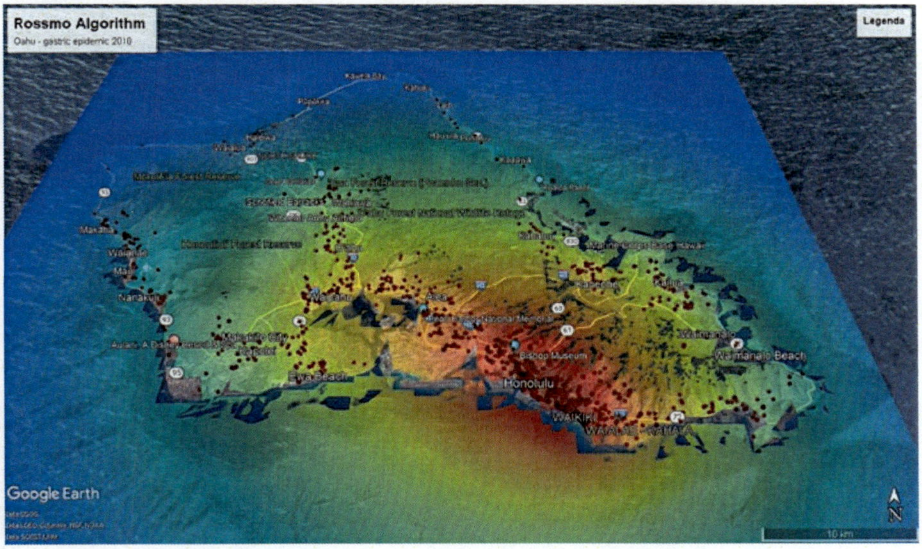

Fig. 6.47 Rossmo algorithm, Oahu food poisoning outbreak

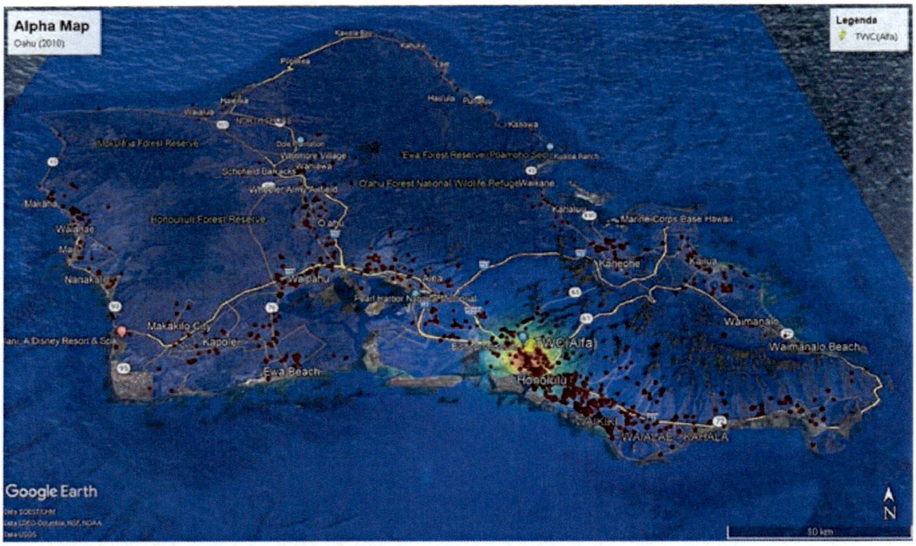

Fig. 6.48 TWC-Alpha Oahu food poisoning

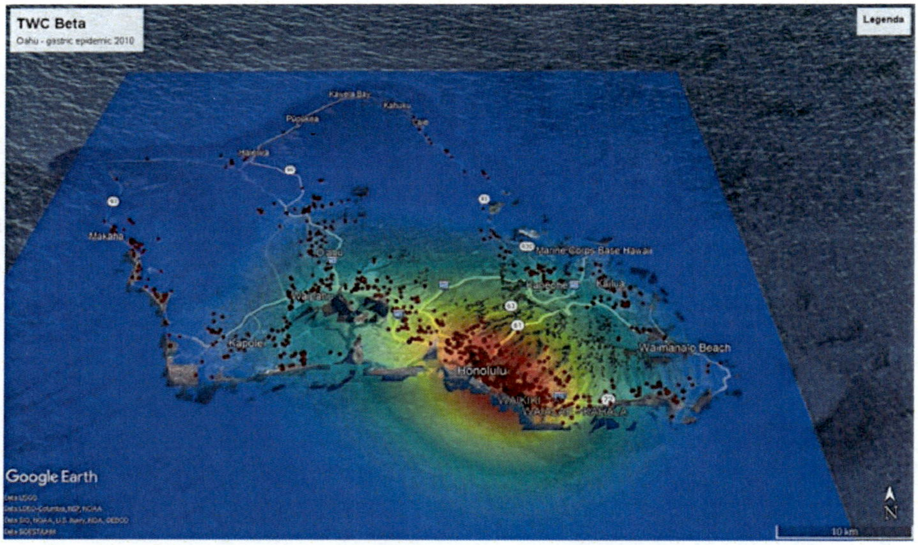

Fig. 6.49 TWC—Beta Oahu food poisoning

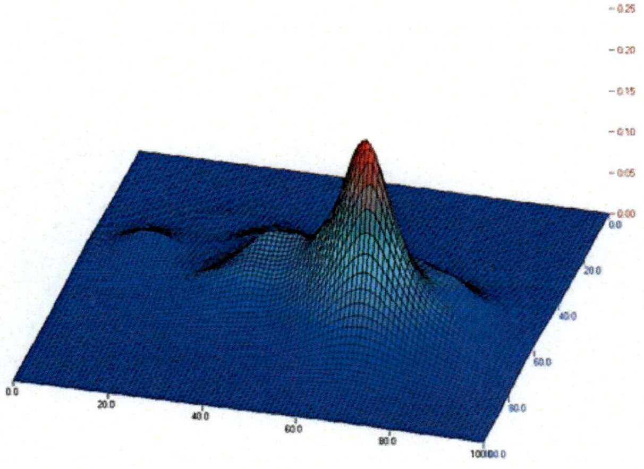

Fig. 6.50 Canter algorithm—Oahu food poisoning

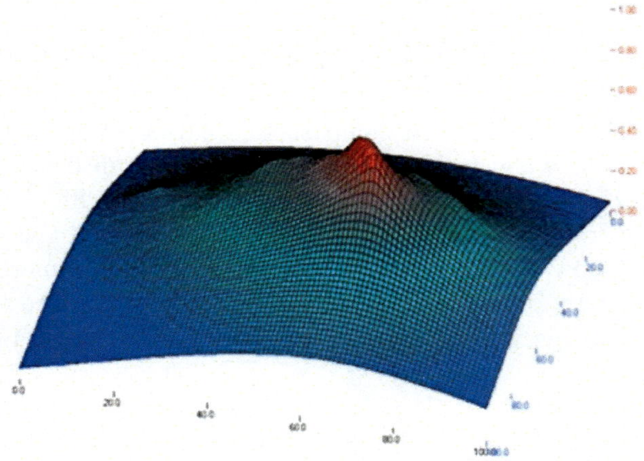

Fig. 6.51 Rossmo algorithm—Oahu food poisoning

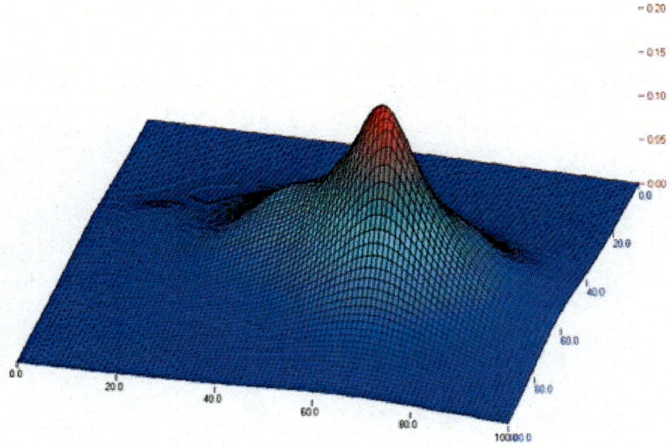

Fig. 6.52 TWC—Alpha Oahu food poisoning

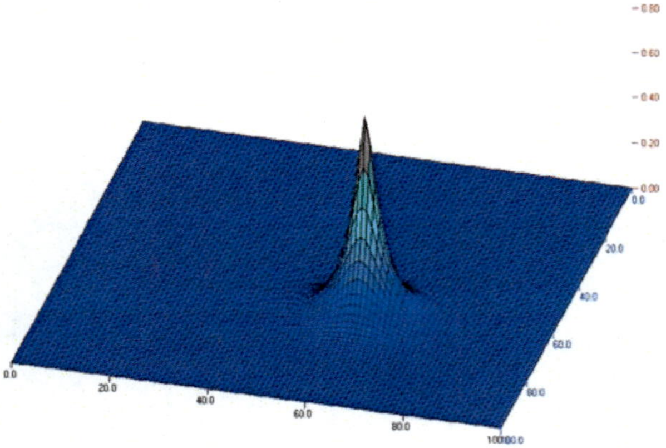

Fig. 6.53 TWC-Beta Oahu food poisoning

6.3.3 *Ebola Epidemic (Africa 2014–2018)*

Ebola epidemic is composed of 82 data points, collected from 2014 and 2018. The following figures shows the data points distribution in central Africa, the scalar fields generated by Canter and Rossmo algorithms. Then two figures show the same provided by TWC-Alpha and TWC-Beta. This application of all the four algorithms show a similar behavior, but TWC-Alpha is more specific and it points out the outbreak in a precise sector that was demonstrated to be less than 50 km away from village where the patient zero was discovered (Figs. 6.54, 6.55, 6.56, 6.57, 6.58,

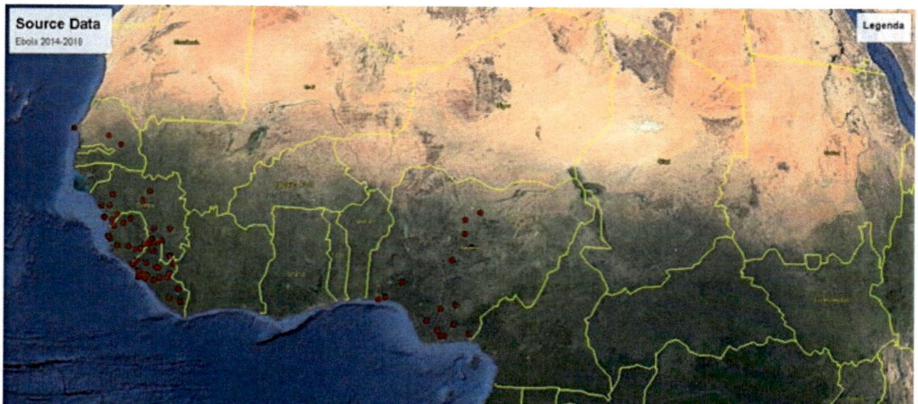

Fig. 6.54 Ebola epidemic data (red dots) from 2014 to 2018

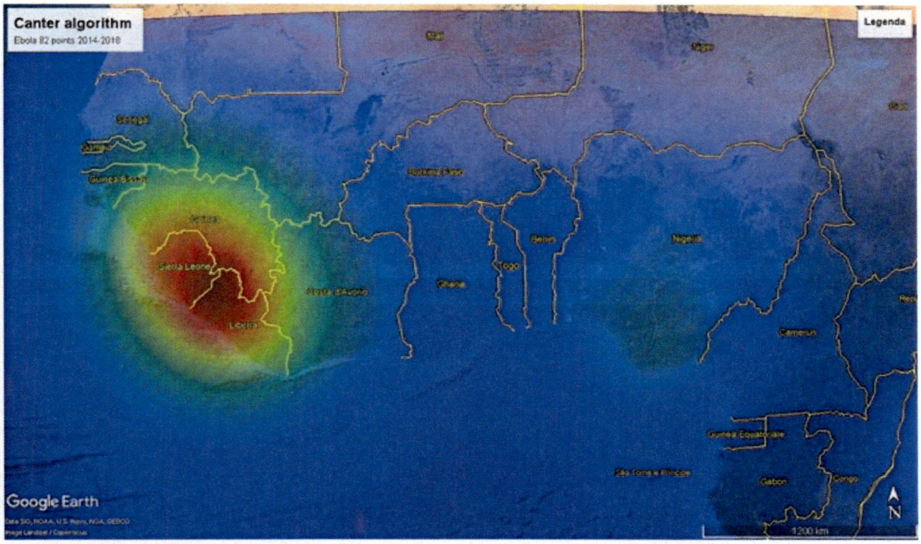

Fig. 6.55 Canter algorithm applied to Ebola

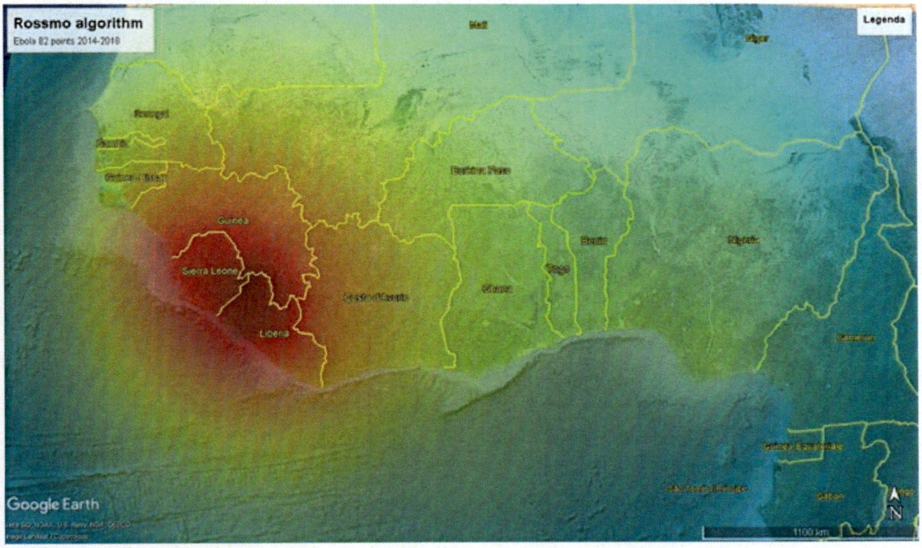

Fig. 6.56 Rossmo algorithm applied to Ebola

6.59, 6.60, 6.61 and 6.62). The next four figures show the scalar fields rebuild by the four algorithms in a $3-D$ surface. This time TWC-Alpha and TWC-Beta maps are very similar, Rossmo map is vague, and Canter map seems to be a mixture between TWC-Alpha and TWC-Beta.

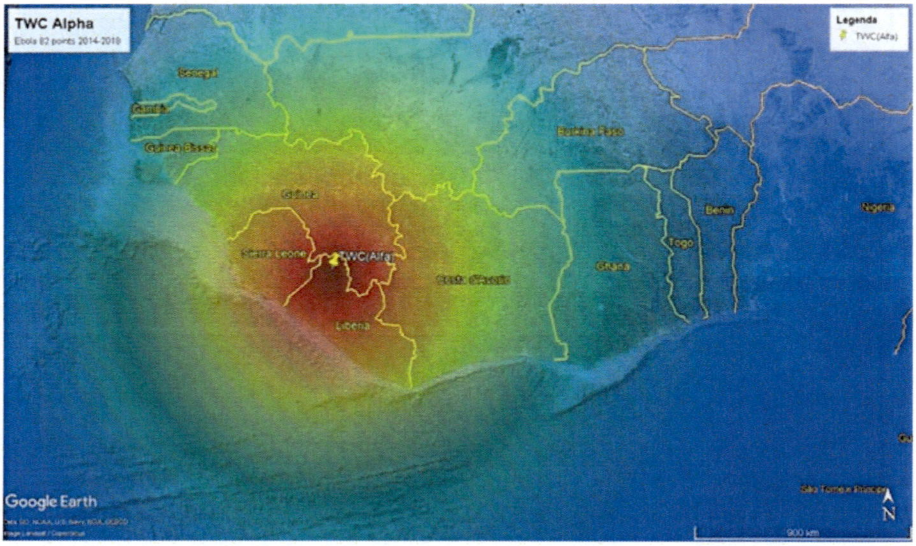

Fig. 6.57 TWC-Alpha applied to Ebola

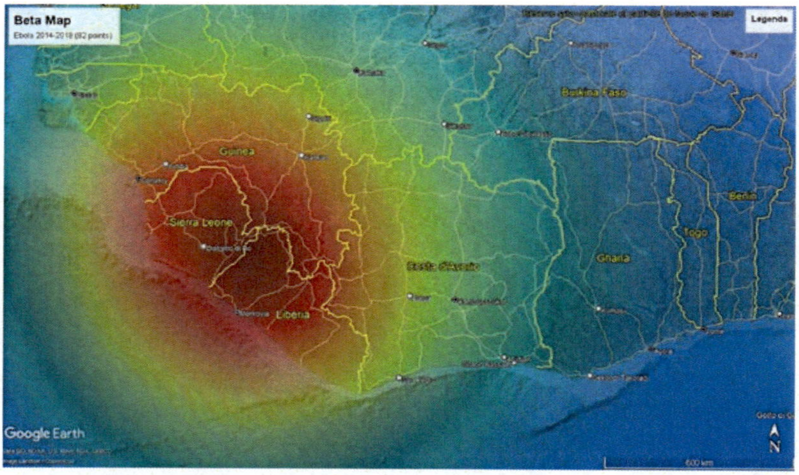

Fig. 6.58 TWC-Beta applied to Ebola

Fig. 6.59 Canter algorithm

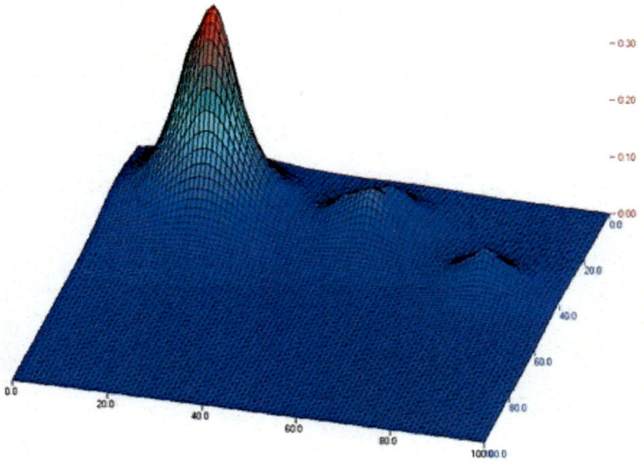

Fig. 6.60 Rossmo algorithm

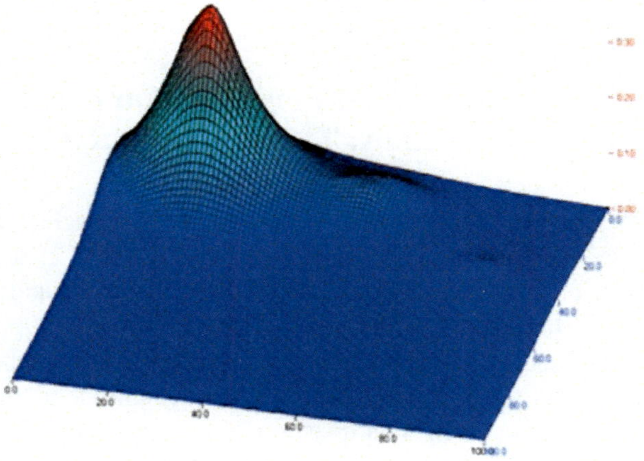

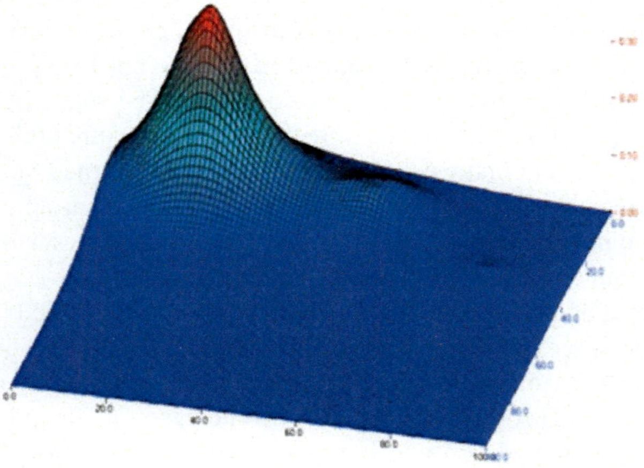

Fig. 6.61 TWC—Alpha

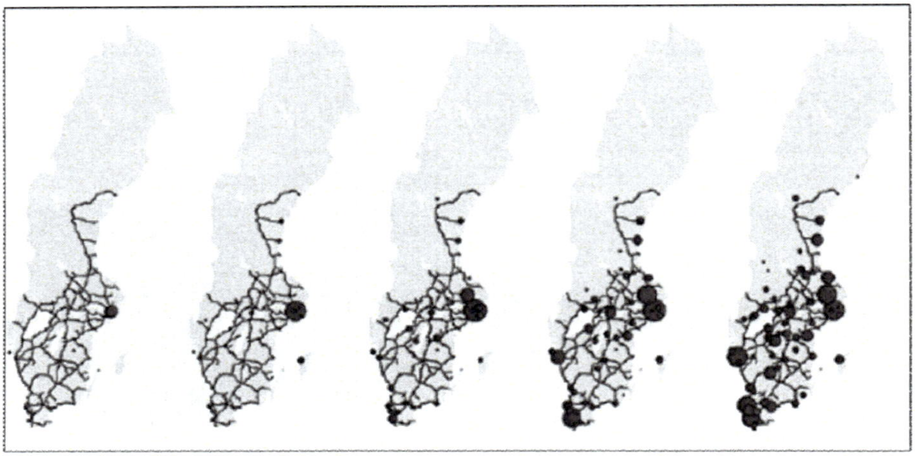

Fig. 6.62 Four-week progression of the Russian Influenza—Sweden 1889

6.4 Why TWC is Powerful Compared to Some Other Algorithms: The Russian Flu in Sweden (1889–1890)

The example presented next, using data from a study of the 1889–90 Russian flu in Sweden, shows how the application of the TWC methodology may help to identify the source of an epidemic outbreak in surveillance analyses. For a complementary study on the sources of outbreaks see also [11]. In 1890, immediately after the outbreak of Russian influenza, all Swedish doctors were asked to provide information about the start and the peak of the epidemic, and the total number of cases in their region, as well as to fill in a questionnaire on the number, sex and age of infected persons in the households they visited. General answers on the epidemic were received from 398 physicians, and data on individual patients were available for more than 32,600 persons. From such answers, a table was compiled and a map was drawn in 1890, indicating when the influenza first appeared at different locations. To support the contagiousness theory, an analysis of the connections between the spatial dynamics of the disease from the onset of its outbreak, and the structure of the railway network was carried out. In the first week of December 1889, it turned out that 12 of the 13 affected places outside Stockholm had railway stations. Linroth [12] demonstrated that, by December 20 of the same year, 82% of reporting places with a railway station and 47% without one had been affected. The dissemination was very fast and the local epidemics developed at a pace that in some cases was described as explosive. Due to the general susceptibility, the short incubation time and the difficulty to detect the very first cases, more evidence was needed to scientifically ascertain that the influenza was indeed contagious. Linroth was however of the opinion that the many individual testimonies describing how the infection was transferred directly from infected persons justified the hypothesis that influenza is a contagious disease.

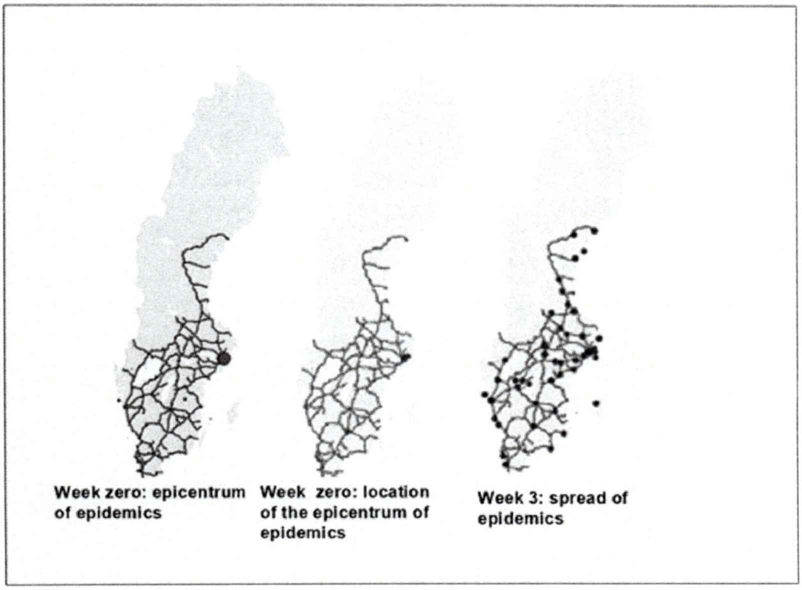

Fig. 6.63 Week 0 versus Week 3 - Swedish Russian Influenza 1889

Linroth's original tables, in a recent GIS study [13], were converted into an Excel format and into dot maps. The figure below, taken from [13], illustrates the progression of the epidemic across time. Dots represent the places where at least one case of infection had been reported to date, starting from the last week of November 1889. The railroad network is also shown (see Fig. 6.63). In Fig. 6.63 the dots sizes are proportional to the number of reported cases. Each map represents one week, starting from the last week of 1889 (first left) (see Fig. 6.64).

The figure below shows a magnification of dot maps at week zero and week three of the Swedish epidemic spread (see Fig. 6.65).

6.4.1 The TWC Analysis of the Swedish Spanish Flu Outbreak

TWC worked on the map of dots of week 3 using the coordinates of the points corresponding to the 44 locations of interest for the outbreak in the third week in order to calculate the appropriate mathematical quantities described in the first part of the article. The coordinates were derived from the dot maps. Figure 6.66 shows the 44 dot map locations obtained from the original coordinates and from which the TWC methodology was implemented.

The source of the epidemic corresponds to a location in between points 25 and 30 in the dot maps. As it is shown in Fig. 6.67, starting from the Euclidean centroid, the

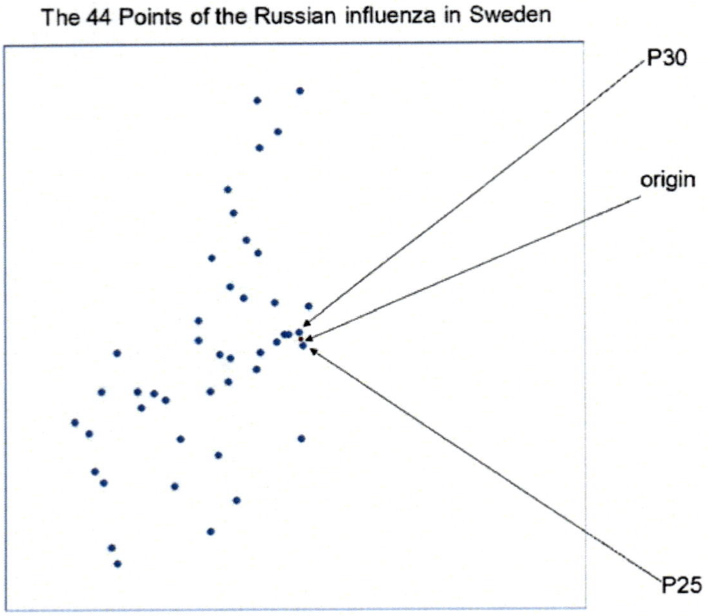

Fig. 6.64 Data for the Swedish Russian Influenza

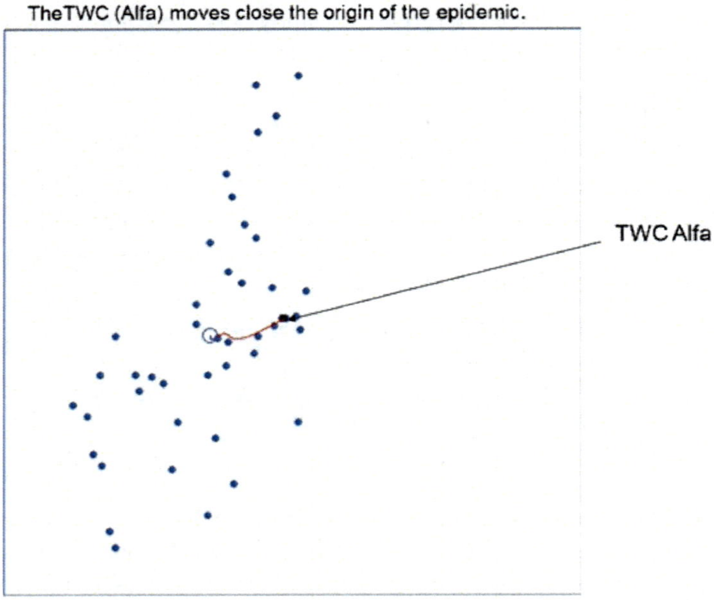

Fig. 6.65 TWC Alpha point versus source point

system positions the TWC alpha point exactly between points 25 and 30, the source of the epidemic spread.

TWC—Beta is depicted in Fig. 6.68, again correctly finding the place of origin of the epidemic. In addition TWC—Beta also provides the probability of its diffusion.

The gradient of the TWC-Beta map yields the likely speed of diffusion of the epidemic from its origin (see Fig. 6.69). It is noteworthy that our emergent map anticipates quite closely the real spread observed in weeks 4 and 5.

The TWC—Gamma map picks even more precisely the origin of the epidemic and the actual probability of its diffusion (see Fig. 6.70).

The gradient of the TWC—Gamma map provides again the likely speed of diffusion of the epidemic from its origin (see Fig. 6.71).

Finally, with TWC—Gamma the possible paths of diffusion of the epidemic are drawn out (Fig. 6.72). It is interesting to notice that such paths are very close to the actual railway network as reported in the maps in [13].

This and other examples, show a different type of complex spatial dynamics, from early detection and its spread, its contagion mechanism. In fact, one might see the contagion dynamics in terms of capture of the infected humans by the viral agent. But the logic of capture in the two cases is basically different in that in the latter the number of predators is extremely high and capture occurs through physical proximity and contact, whereas in the former predators are few and capture occurs through careful intelligence and a dose of good luck. Nevertheless, we also find that

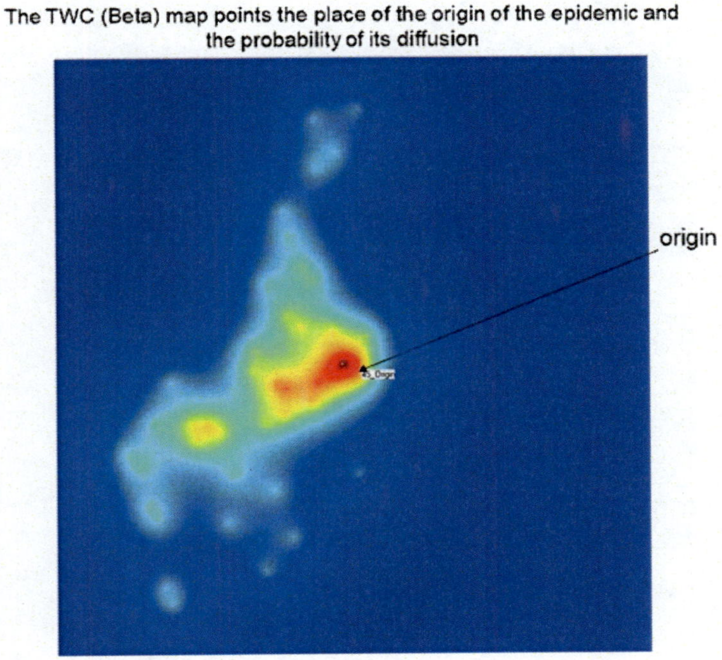

Fig. 6.66 TWC—Beta map Swedish Russian Influenza

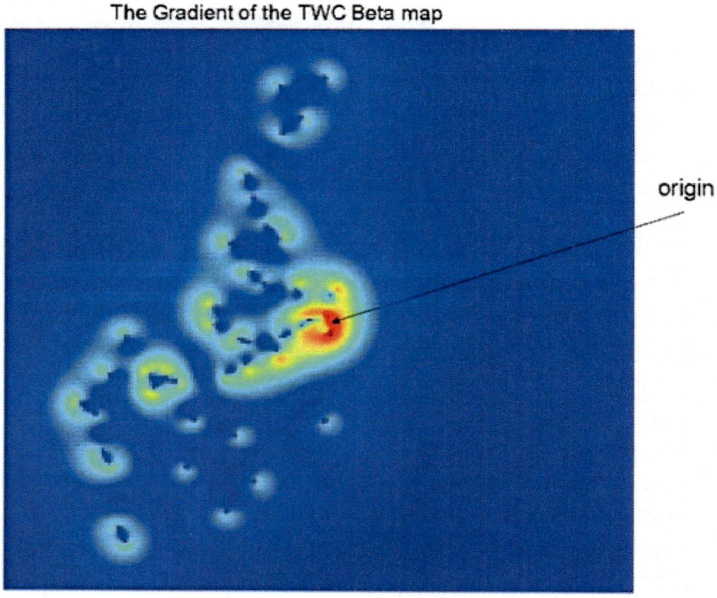

Fig. 6.67 Gradient of the TWC—Beta map

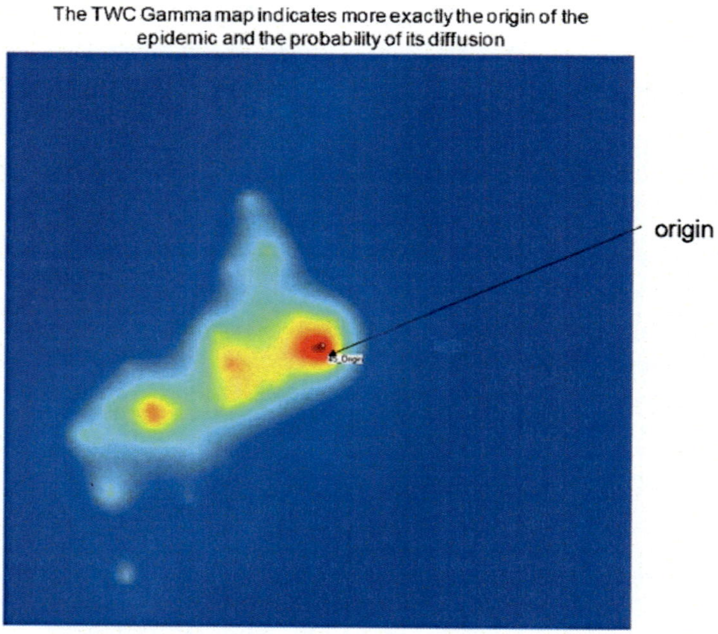

Fig. 6.68 TWC—Gamma Swedish Russian Influenza

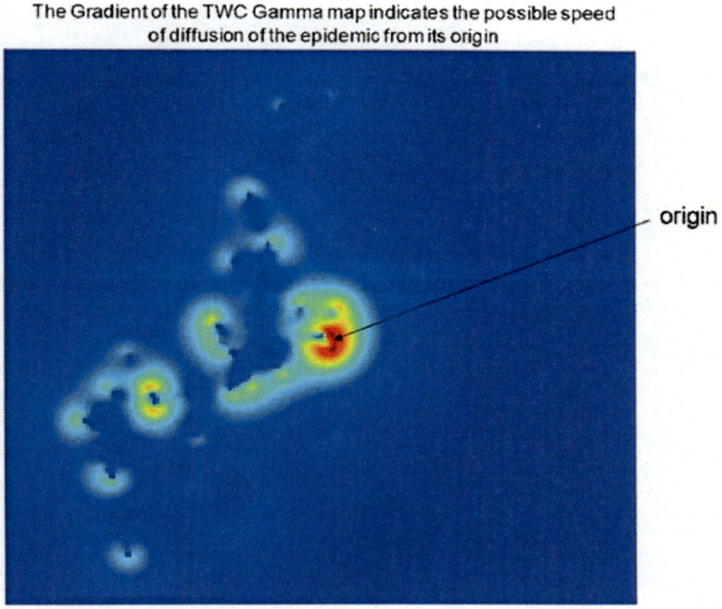

Fig. 6.69 Gradient map of TWC—Beta Swedish Russian Influenza

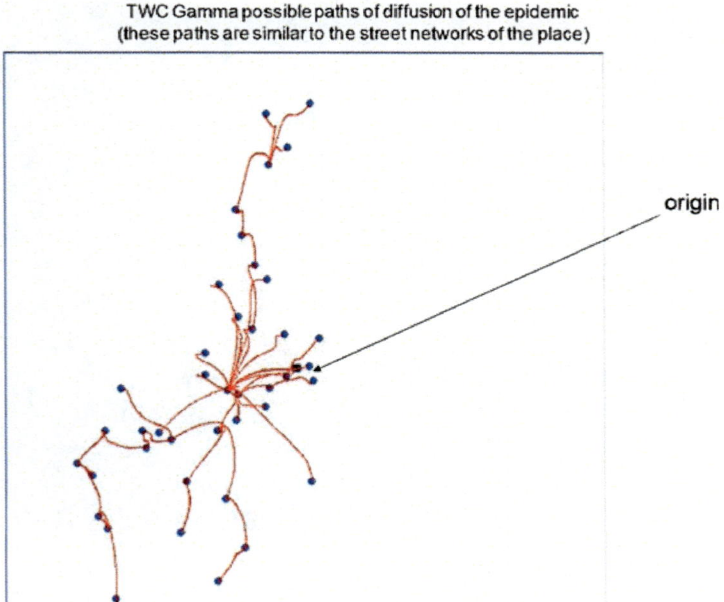

Fig. 6.70 TWC—Gamma paths: Swedish Russian Influenza

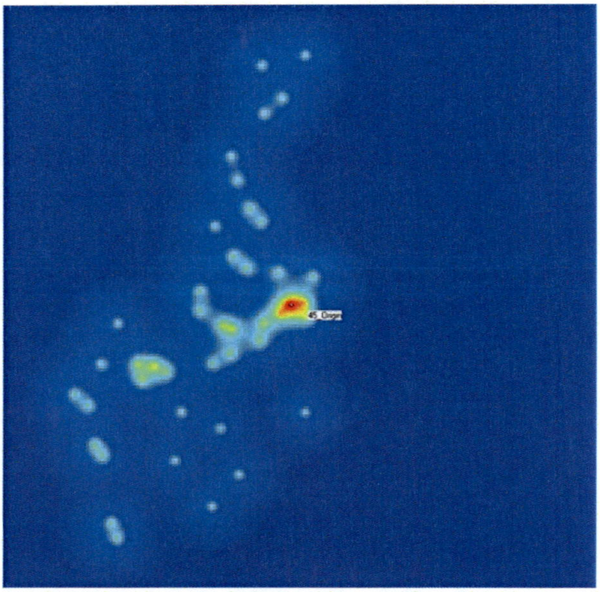

Fig. 6.71 Canter algorithm—Swedish Russian Influenza

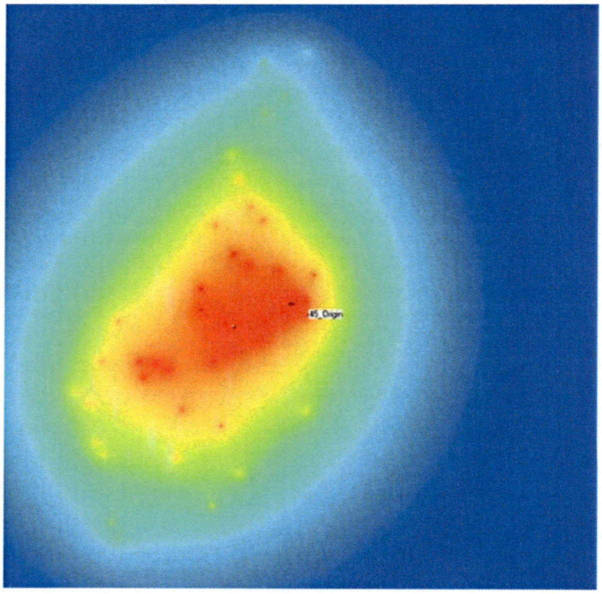

Fig. 6.72 Rossmo algorithm: Swedish Russian Influenza

this entirely different kind of spatial phenomenon may be addressed through the same set of tools, which provide a rather accurate reconstruction of the underlying structural forces at play. Again, this is a case where thinking of the phenomenon in ‘relational’ terms and trying to reconstruct its implicit semantics makes sense, and this time in a very intuitive way. Each case of the contagion clearly determines an area of influence (morbidity) that drives the appearance of further cases, greater or less so depending on a number of local circumstances (size of the family, type of housing, pre-existence of other diseases, etcetera). In addition the long-range interactions determined by the railway system are also mirrored in the results with a high level of accuracy. This is a clear illustration of the global character of the relational concept on which the TWC methodology is based.

6.4.2 Comparison TWC with Other Algorithms

The case of Russian influenza in Sweden is pretty well known. In this case, we know the exact location of the outbreak and of its dynamics, and thus we can carry out any kind of benchmark comparison test. Specifically, we can compare the performance of TWC algorithms with that of the other algorithms.

The algorithms we have selected for this benchmark are the following seven.

1. Two simple algorithms that are only able to work upon a minimization criterion with a specific cost function:
   (a) The Anchor Point;
   (b) The Center of Mass.
2. Five (some well known, some new) algorithms that estimate the probability of the outbreak for each point of the convex hull of the map:
   (a) The Rossmo Algorithm;
   (b) The Canter Algorithm;
   (c) The NES Algorithm;
   (d) The LVM Algorithm;
   (e) The MexProb Algorithm.
3. TWC algorithms that are selected are three:
   (a) i. The TWC-Alpha point;
       ii. The TWC—Beta map;
       iii. The TWC-Gamma map.

We also considered various criteria to compare the performance of the algorithms.

1. **Percent Error**: The percent distance of the estimated point and/or of the estimated peak of the map generated by the algorithm from the real outbreak. We have scaled this measure in relation to the maximum distance among the assigned points.

2. **Sensitivity:** The probability that each algorithm assigns in its characteristic map to the point where the real outbreak occurred (only for the algorithms whose output is a probabilistic map).
3. **Specificity**: If we divide the probability distribution into 9 equivalent bins from 0 to 1, we have 9 classes of probability; the percent of points of all the classes of the map where the real outbreak point is not located defines the specificity of the algorithm. The table and the figures below show a general convergence of the more complex algorithms toward the real outbreak point.

		Distance from the Outbreak		
Algorithms	Type of Output	**Accuracy**	Sensitivity	Specificity
TWC—Gamma	Probability Map (Peak)	1.67%	90.13%	99.89%
TWC—Alpha	Point	3.13%		
Canter	Probability Map (Peak)	3.81%	58.86%	99.83%
TWC—Beta	Probability Map (Peak)	4.55%	77.59%	99.50%
NES	Probability Map (Peak)	4.55%	79.08%	99.46%
MexProb	Probability Map (Peak)	6.73%	82.69%	99.32%
LVM	Probability Map (Peak)	6.77%	84.57%	98.14%
Anchor Point:	Point	15.90%		
Mean Center:	Point	17.44%		
Rossmo	Probability Map (Peak)	19.42%	80.18%	96.36%

Table-Comparison of methods

Some observations and remarks are in order.

1. TWC—Alpha is the best estimation among the algorithms with point-wise output, and it is the second best in the general ranking.
2. TWC—Gamma peak points out the real outbreak with a very consistent accuracy, high sensitivity and very high specificity.
3. Canter algorithm is quite accurate but its sensitivity is low. Rossmo algorithm is not accurate and its specificity is lower than the others.
4. The other algorithms are positioned in middle range of performance: A reasonable accuracy, an acceptable sensitivity and a high specificity.

The following Figures, (see Figs. 6.73, 6.74, 6.75 and 6.76), depict the results, with the following conventions:

- The colors from dark blue to dark red indicate the estimation of the operation point location from 0.0 to 1.0;
- The name of the algorithm is at the top of the figure;
- The white point represents the peak of the map;
- The label "45_Origin" represents the real outbreak.

The anchor point, or center of mass, is found in Fig. 6.67, the open small blue circle that is at—the end of the red alpha path.

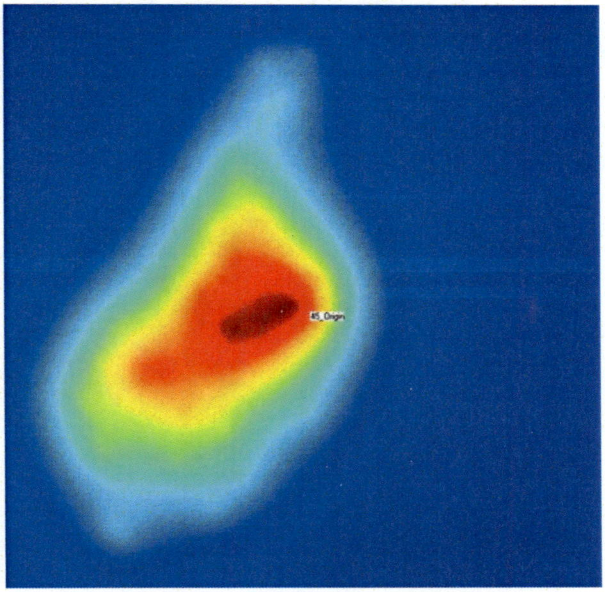

Fig. 6.73 Anchor algorithm: Swedish Russian Influenza

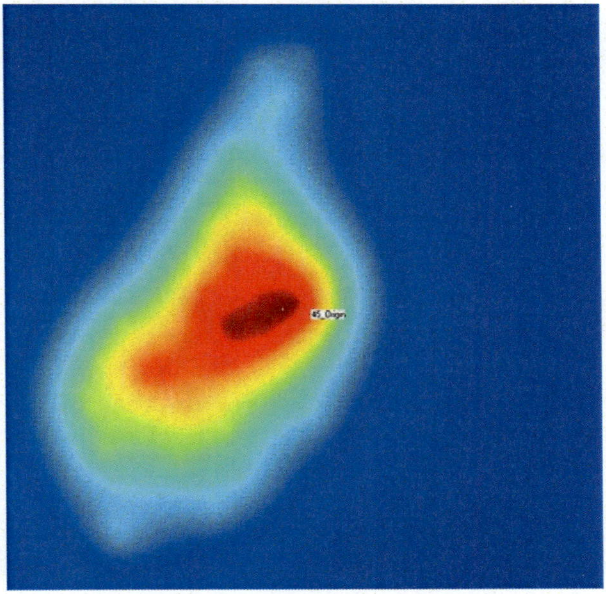

Fig. 6.74 NES algorithm: Swedish Russian Influenza

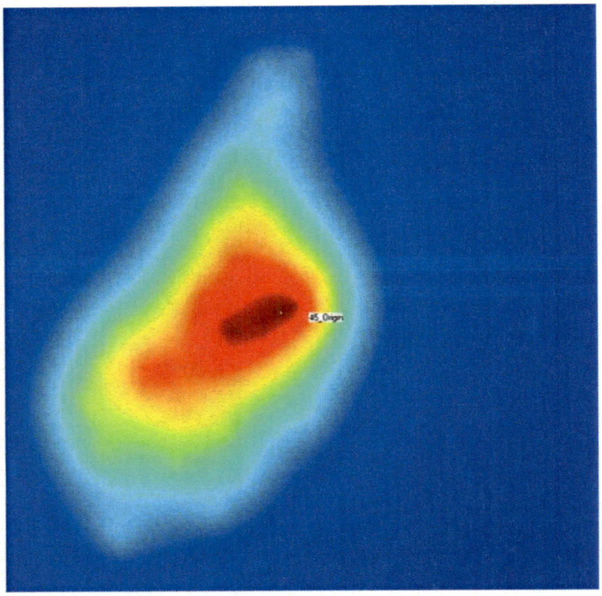

Fig. 6.75 Likelihood Variance Maximization Algorithm LVM Algorithm

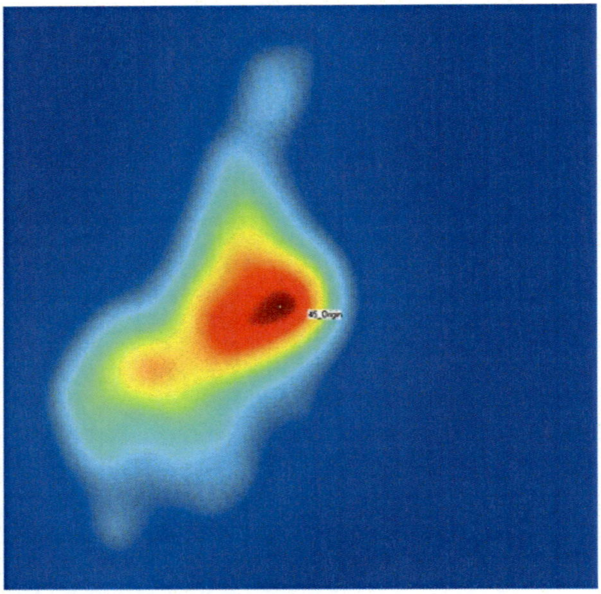

Fig. 6.76 Mexican probability algorith: Swedish Russian Influenza

6.5 Strengths and Weakness of TWC

The applications and comparisons are just are a little part of the real cases that have been analyzed using TWC. Further applications are found in the literature. What is presented in this monograph represents a sample of the cases. For each of the cases shown, the location identified was an approximation of the real solution. In other words, TWC generates a solution that is a region (the "hot" spots) that corresponds to the sources. Most of the applications had limited (few) data and in some cases imprecise data. It remains to test whether with good data or more precise data, a greater precision of the locations ensue. "Good data" means that:

1. All the data points (events) in fact belong to the same process;
2. The data points represent a good statistical representation of the real process;
3. The source of the process is inside the convex hull of the assigned data points.

The TWC algorithm risks failing when these conditions are not present in the data. But when all these conditions are more or less satisfied, the results are good. TWC tries to find out what the spacial distribution of points are trying to tell us.

References

1. M. Buscema, G. Monaghan, P. Richards P, (eds.), *CDTD, Central Drug Trafficking Database, in MPS Drugs Strategy 2007–2010 and Delivery Plan", Appendix 1, Chapter 6* (London Metropolitan Police, London, 2007)
2. M. Buscema, *Squashing Theory and Contractive Map Network, Semeion Technical Paper #32* (Rome, 2007)
3. M. Buscema, E. Grossi, The semantic connectivity map: an adapting self-organizing knowledge discovery method in data bases. Experience in gastro-oesophageal reflux disease. Int. J. Data Min. Bioinform. **2**, 362–404 (2008)
4. M. Buscema, P.L. Sacco, *Auto-contractive Maps, the H Function and the Maximally Regular Graph (MRG): A New Methodology for Data Mining, mimeo* (Semeion, Rome, 2008)
5. M. Buscema, C. Helgason, E. Grossi, *Auto contractive maps, h function and maximally regular graph: theory and applications, special session on artificial adaptive systems in medicine: applications to the real world, in NAFIPS 2008* (IEEE, New York, 2008)
6. M. Buscema, M.A. Zeydabadi, W. Lodwick, A. Nde Nembot, A. Bronstein, F. Newman, Analysis of the Ebola outbreak in 2014 and 2018 in West Africa and Congo by using artificial adaptive systems, in *Applied Artificial Intelligence* (2020). https://doi.org/10.1080/08839514.2020.1747770
7. M.A. Zeydabadi, M. Buscema, W. Lodwick, G. Massini, F. Della Torre, F. Newman, Analysis of COVID-19 pandemic in USA, using topological weighted centroid, in *Computers in Biology and Medicine*, vol. 136. https://doi.org/10.1016/j.compbiomed.2021.104670
8. M. Buscema, E. Grossi, A. Bronstein, W. Lodwick, M.A. Zeydabadi, R. Benzi, F. Newman, A new algorithm for identifying possible epidemic sources with application to the German Escherichia Coli outbreak. ISPRS Int. J. Geo-Inf. **2**, 155–200 (2013). https://doi.org/10.3390/ijgi2010155
9. A.C. Bronstein, M. Buscema, A. Esfahani, W. Lodwick, E. Grossi, Locating the source of public health events using intelligent adaptive systems: 2011 United States listeriosis outbreak linked to whole cantaloupes. Clin. Toxicol. **51**(7), 625–626 (2013). (2013 Annual Meeting of the North American Clinical Toxicology (NACCT): Platforms 1-20; Posters 21-316)

10. M. Buscema, E. Grossi, M. Breda, T. Jefferson, Outbreaks source: a new mathematical approach to identify their possible location. Phys. A **388**, 4736–4762 (2009)
11. an example of geographical information system analysis, L. Skog L, H. Hauska, A. Linde, The Russian Influenza in Sweden 1889–1890. Eurosurveillance **13**, 1–7 (2008)
12. K. Linroth, Influensan i Sverige 1889-1890 enligt iakttagelser af landets lakäre, in *Del. I: Influensan i epidemiologiskt hänseende. Svenska Läkaresällskapets Nya Handlingar. Serie III* (1890), pp. 1–92
13. M. Buscema, S. Terzi, A new evolutionary Approach to topographic mapping, in *Proceedings of the 7th WSEAS International Conference on Evolutionary Computing* (Cavtat, Croatia, 2006), pp. 12-19

Chapter 7
Appendices—MATLAB Programs

These appendices contain four of TWC the algorithms (alpha, beta, gamma, theta) written in MATLAB code. The MATLAB code was used to generate the maps associated with the illustrative examples Chap. 4. The code is designed to be used with the data tables in the book and those that the reader might wish to use to test out the approaches associated with some of the TWC maps. The correctness of the code is not guaranteed though it has been used to generate the examples that are in the monograph, not the applications. That is, the code presented here was used to produce the TWC maps associated with the 10 and 11 point examples (4.1), (4.2). To run the code, MATLAB needs the latitude/longitude of the users' data.

7.1 TWC-Alpha

```
%TWC_alpha__Code
   clear all
   % A is matrix of x and y coordinates.
   A = importdata('xy.dat');
   x=A(:,1);
   y=A(:,2);
   figure(1)
   plot(x,y,'*b'),xlabel('x'),ylabel('y'), axis equal, grid on
   figure(2)
   plot(x,y,'*b'),xlabel('x'),ylabel('y'), axis equal
   hold on
   N=length(x);
   %dij
   for i=1:N
       for j=1:N
```


M. Buscema et al., *The Topological Weighted Centroid: A New Vision of Geographic Profiling*, Studies in Computational Intelligence 1095,
https://doi.org/10.1007/978-3-031-28901-9_7

```
            d(i,j)=sqrt((x(i)-x(j))^2+(y(i)-y(j))^2);
        end
    end
    for i=1:N
        for j=1:N
            d_bar(i,j)=0;
            if i~=j
                d_bar(i,j)=sum(d(i,:))-d(i,j);
            end
        end
    end
    d_bar=d_bar/(N-2);
    del_d=d-d_bar;
    D=max(max(d_bar));
    %w
    for i=1:N
        w(i)=0;
        for j=1:N
            if i~=j
                w(i)=w(i)+exp(-d_bar(i,j)/D);
            end
        end
    end
    w=w/(N-1);
    del_alpha=input('del_alpha= '); %for example: del_alpha= 0.01;
    n_max=input('n_max= '); %for example= 10000;
    %w(alpha)
    for n=1:n_max
        alpha(n)=(n-1)*del_alpha;
        for i=1:N
            w_alpha(n,i)=0;
            for j=1:N
                if i~=j
                    w_alpha(n,i)=w_alpha(n,i)+exp(-d_bar(i,j)/D*alpha(n));
                end
            end
        end
    end
    w_alpha=w_alpha/(N-1);%TWC_alpha
    for n=1:n_max
        s_w_alpha(n)=0;
        TWCx(n)=0;
        TWCy(n)=0;
        for i=1:N
            s_w_alpha(n)=s_w_alpha(n)+w_alpha(n,i);
```

```
        TWCx(n)=TWCx(n)+w_alpha(n,i)*x(i);
        TWCy(n)=TWCy(n)+w_alpha(n,i)*y(i);
    end
    TWCx(n)=TWCx(n)/s_w_alpha(n);
    TWCy(n)=TWCy(n)/s_w_alpha(n);
    figure(2)
    plot(TWCx(1),TWCy(1),'*r',TWCx(n),TWCy(n),'.r'), grid on
    drawnow
end
```

7.2 TWC-Beta

```
% TWC_beta_scalar_Code
  clear all
  % A is matrix of x and y coordinates.
  A = importdata('xy.dat');
  x=A(:,1);
  y=A(:,2);
  figure(1)
  plot(x,y,'*b'),xlabel('x'),ylabel('y'), axis equal
  hold on
  N=length(x);
  %dij
  %dij
  for i=1:N
      for j=1:N
          d(i,j)=sqrt((x(i)-x(j))^2+(y(i)-y(j))^2);
      end
  end
  for i=1:N
      for j=1:N
          d_bar(i,j)=0;
          if i~=j
              d_bar(i,j)=sum(d(i,:))-d(i,j);
          end
      end
  end
  d_bar/(N-2);
  del_d_n=d-d_bar;
  D=max(max(d_bar));
  %w
  del_beta=input('del_beta= '); % For example del_beta=0.01;
  n_max_b=input('n_max_b= '); %For example n_max_b=10000;
```

```
%w(alpha)
for n=1:n_max_b
    beta(n)=(n-1)*del_beta;
    for i=1:N
        w_beta(n,i)=0;
        for j=1:N
            w_beta(n,i)=w_beta(n,i)+exp(-d_bar(i,j)/D*beta(n));
        end
    end
end
w_beta=w_beta/(N-1); %w_beta=w_beta/N;
%TWC_beta
for n=1:n_max_b
    TWC_bet_x(n)=0;
    TWC_bet_y(n)=0;
    Sum_w_beta(n)=0;
    for i=1:N
        TWC_bet_x(n)=TWC_bet_x(n)+w_beta(n,i)*x(i);
        TWC_bet_y(n)=TWC_bet_y(n)+w_beta(n,i)*y(i);
        Sum_w_beta(n)=Sum_w_beta(n)+w_beta(n,i);
    end
    TWC_bet_x(n)=TWC_bet_x(n)/Sum_w_beta(n);
    TWC_bet_y(n)=TWC_bet_y(n)/Sum_w_beta(n);
    CM_x=TWC_bet_x(1);
    CM_y=TWC_bet_y(1);
    d_TWC_CM(n)=sqrt((TWC_bet_x(n)-CM_x)^2+(TWC_bet_y(n)-CM_y)^2);
    figure(1)
    plot(TWC_bet_x(n),TWC_bet_y(n),'.r')
    drawnow
end
for n=2:n_max_b-1
    if d_TWC_CM(n)>= d_TWC_CM(n-1)& d_TWC_CM(n)>=
d_TWC_CM(n+1)
        n_bet=n;
    end
end
d_TWC_CM_bet=max(sqrt((TWC_bet_x-CM_x).^2+(TWC_bet_y-CM_y).^2));
d_TWC_CM(n_bet);
bet_str=beta(n_bet);
%%%%%%%%%%%%%%%
del=input('del= '); % For example del=1;
x_p=min(x)-30*del:del:max(x)+30*del;
y_p=min(y)-30*del:del:max(y)+30*del;
for k1=1:length(x_p)
    for j1=1:length(y_p)
```

```
            for k2=1:length(x_p)
                for j2=1:length(y_p)
                    n(k1,j1,k2,j2)=sqrt((x_p(k1)-x_p(k2))^2+(y_p(j1)-y_p(j2))^2);
                end
            end
        end
    end
    for i=1:N
        F(i)=0;
        for j=1:N
            F(i)=F(i)+exp(-d_bar(i,j)/D)/(N-1);
        end
    end
    for k1=1:length(x_p)
        for j1=1:length(y_p)
            b(k1,j1)=0;
            for i=1:N
                n(k1,j1,i)=sqrt((x_p(k1)-x(i))^2+(y_p(j1)-y(i))^2);
                b(k1,j1)=b(k1,j1)+F(i)*exp(-n(k1,j1,i)*bet_str/D)/N;
            end
        end
    end
    %%%%%%%%%%%%%%%%%%%%%%
    for i=1:length(x_p)
        for j=1:length(y_p)
            xx(i,j)=x_p(i);
            yy(i,j)=y_p(j);
        end
    end
    figure(3)
    contourf(xx,yy,b),title('TWC\beta')
    figure(4)
    surface(x_p,y_p,b'),title('TWC\beta')
    figure(5)
    surf(x_p,y_p,b'),title('TWC\beta')
```

7.3 TWC-Gamma

```
%TWC_gamma_Code
    clear all
    % A is matrix of x and y coordinates.
    A = importdata('xy.dat');
    x=A(:,1);
```

```
y=A(:,2);
figure(1)
plot(x,y,'*b'),xlabel('x'),ylabel('y'), axis equal
hold on
N=length(x);
%dij
%dij
    for i=1:N
        for j=1:N
            d(i,j)=sqrt((x(i)-x(j))^2+(y(i)-y(j))^2);
        end
    end
D=max(max(d));
del_gam=input('del_gam= '); %For example: del_gam=0.01;
n_max=input('n_max= '); %For example: n_max=500;
for n=1:n_max
    gamma(n)=(n-1)*del_gam;
    for i=1:N
        TWC_gam_x(i,n)=0;
        TWC_gam_y(i,n)=0;
        Z(i,n)=0;
        for j=1:N
            P_gam(i,j,n)=exp(-d(i,j)*gamma(n)/D);
            TWC_gam_x(i,n)=TWC_gam_x(i,n)+P_gam(i,j,n)*x(j);
            TWC_gam_y(i,n)=TWC_gam_y(i,n)+P_gam(i,j,n)*y(j);
            Z(i,n)=Z(i,n)+P_gam(i,j,n);
        end
        TWC_gam_x(i,n)=TWC_gam_x(i,n)/Z(i,n);
        TWC_gam_y(i,n)=TWC_gam_y(i,n)/Z(i,n);
        figure(1)
plot(TWC_gam_x(i,1),TWC_gam_y(i,1),'*r',TWC_gam_x(i,n),TWC_gam_y
(i,n),'.g'),grid on
drawnow
    end
end
```

7.4 TWC-Theta

```
%TWC_theta_Code
clear all
% A is matrix of x and y coordinates.
A = importdata('xy.dat');
x=A(:,1);
```

```
y=A(:,2);
figure(1)
plot(x,y,’*b’),xlabel(’x’),ylabel(’y’), axis equal
hold on
N=length(x);
%dij
%dij
for i=1:N
    for j=1:N
        d(i,j)=sqrt((x(i)-x(j))ˆ2+(y(i)-y(j))ˆ2);
    end
end
Dmax=max(max(d));
for i=1:N
    for j=1:N
        d_bar(i,j)=0;
        if i˜=j
            d_bar(i,j)=sum(d(i,:))-d(i,j);
        end
    end
end
d_bar/(N-2);
del_d_n=d-d_bar;
D=max(max(d_bar));
%w
del_beta=input(’del_beta= ’); %For example: del_beta=0.01;
n_max_b=input(’n_max_b= ’); %For example: n_max_b=10000;
%w(alpha)
for n=1:n_max_b
    beta(n)=(n-1)*del_beta;
    for i=1:N
        w_beta(n,i)=0;
        for j=1:N
            w_beta(n,i)=w_beta(n,i)+exp(-d_bar(i,j)/D*beta(n));
        end
    end
end
w_beta=w_beta/(N-1); %w_beta=w_beta/N;
%TWC_beta
for n=1:n_max_b
    TWC_bet_x(n)=0;
    TWC_bet_y(n)=0;
    Sum_w_beta(n)=0;
    for i=1:N
        TWC_bet_x(n)=TWC_bet_x(n)+w_beta(n,i)*x(i);
```

```
            TWC_bet_y(n)=TWC_bet_y(n)+w_beta(n,i)*y(i);
            Sum_w_beta(n)=Sum_w_beta(n)+w_beta(n,i);
        end
        TWC_bet_x(n)=TWC_bet_x(n)/Sum_w_beta(n);
        TWC_bet_y(n)=TWC_bet_y(n)/Sum_w_beta(n);
        CM_x=TWC_bet_x(1);
        CM_y=TWC_bet_y(1);
        d_TWC_CM(n)=sqrt((TWC_bet_x(n)-CM_x)^2+(TWC_bet_y(n)-CM_y)^2);
        figure(1)
        plot(TWC_bet_x(n),TWC_bet_y(n),'.r')
        drawnow
    end
    for n=2:n_max_b
        if d_TWC_CM(n)>= d_TWC_CM(n-1)& d_TWC_CM(n)>=
d_TWC_CM(n+1)
            n_bet=n
        end
    end
    d_TWC_CM_bet=max(sqrt((TWC_bet_x-CM_x).^2+(TWC_bet_y-CM_y).^2));
    d_TWC_CM(n_bet);
    bet_str=beta(n_bet);
    %%%%%%%%%%%%%%%
    del_gam=input('del_gam= '); %For example: del_gam=0.01;
    n_max=input('n_max= '); %For example: n_max=10000;
    for n=1:n_max
        gamma(n)=(n-1)*del_gam;
        for i=1:N
            TWC_gam_x(i,n)=0;
            TWC_gam_y(i,n)=0;
            Z(i,n)=0;
            for j=1:N
                P_gam(i,j,n)=exp(-d(i,j)*gamma(n)/Dmax);
                TWC_gam_x(i,n)=TWC_gam_x(i,n)+P_gam(i,j,n)*x(j);
                TWC_gam_y(i,n)=TWC_gam_y(i,n)+P_gam(i,j,n)*y(j);
                Z(i,n)=Z(i,n)+P_gam(i,j,n);
            end
            TWC_gam_x(i,n)=TWC_gam_x(i,n)/Z(i,n);
            TWC_gam_y(i,n)=TWC_gam_y(i,n)/Z(i,n);
            figure(1)
            plot(TWC_gam_x(i,1),TWC_gam_y(i,1),'*r',TWC_gam_x(i,n),
TWC_gam_y(i,n),'.g'),grid on
            drawnow
        end
    end
```

```
%%%%%%%%%%%
del=input('del= '); %For example: del=0.1;
x_p=min(x)-5*del:del:max(x)+20*del;
y_p=min(y)-5*del:del:max(y)+20*del;
for i=1:length(x_p)
    for j=1:length(y_p)
    c(i,j)=0;
        for n=1:N
            for t=1:n_max
                m(i,j,n,t)=sqrt((x_p(i)-TWC_gam_x(n,t))^2+(y_p(j)-
TWC_gam_y(n,t))^2);
                c(i,j)=c(i,j)+exp(-m(i,j,n,t)*bet_str/Dmax)/n_max;
            end
        end
    end
end
for i=1:length(x_p)
    for j=1:length(y_p)
        xx(i,j)=x_p(i);
        yy(i,j)=y_p(j);
    end
end
figure(2)
contourf(xx,yy,c),title('TWC\gamma')
figure(3)
surface(x_p,y_p,c'),title('TWC\gamma')
figure(4)
surf(x_p,y_p,c'),title('TWC\gamma')
t_max=n_max
i_thet=1;
TWC_ijt_thet_x=[];
TWC_ijt_thet_y=[];
for t=1:t_max
    TWC_ij_thet_x=[];
    TWC_ij_thet_y=[];
    for i=1:N
        for j=1:N
            TWC_thet_x(i,j,t)=x(j)+TWC_gam_x(i,t)-TWC_gam_x(j,t);
            TWC_thet_y(i,j,t)=y(j)+TWC_gam_y(i,t)-TWC_gam_y(j,t);
        end
    end
end
for i=1:N
    for j=1:N
        thet_ij(i,j)=0;
```

```
            for t=1:t_max-1
                thet_ij(i,j)=(TWC_thet_x(i,j,t+1)-TWC_thet_x(i,j,t))^2-
(TWC_thet_y(i,j,t+1)-TWC_thet_y(i,j,t))^2;
            end
        end
    end
    max_thet=max(max(thet_ij));
    thet_ij_tilde=max_thet-thet_ij;
    for i=1:N
        sum_thet(i)=0;
        for k=1:N
            sum_thet(i)=sum_thet(i)+thet_ij_tilde(i,k);
        end
    end
    for i=1:N
        for j=1:N
            P_thet(i,j)=thet_ij_tilde(i,j)/sum_thet(i);
        end
    end
```

7.5 TWC-Theta Minimal Spanning Tree

```
%TWC_theta_MST_Code
    clear all
    % A is matrix of x and y coordinates.
    A = importdata('xy.dat');
    x=A(:,1);
    y=A(:,2);
    figure(1)
    plot(x,y,'*b'),xlabel('x'),ylabel('y'), axis equal
    hold on
    N=length(x);
    %dij
    %dij
    for i=1:N
        for j=1:N
            d(i,j)=sqrt((x(i)-x(j))^2+(y(i)-y(j))^2);
        end
    end
    Dmax=max(max(d));
    for i=1:N
        for j=1:N
            d_bar(i,j)=0;
```

```
            if i~=j
                d_bar(i,j)=sum(d(i,:))-d(i,j);
            end
        endend
    d_bar/(N-2);
    del_d_n=d-d_bar;
    D=max(max(d_bar));
    %w
    del_beta=input('del_beta= '); %For example: del_beta=0.01
    n_max_b=input('n_max_b= '); %For example: n_max_b=10000
    %w(alpha)
    for n=1:n_max_b
        beta(n)=(n-1)*del_beta;
        for i=1:N
            w_beta(n,i)=0;
            for j=1:N
                w_beta(n,i)=w_beta(n,i)+exp(-d_bar(i,j)/D*beta(n));
            end
        end
    end
    w_beta=w_beta/(N-1); %w_beta=w_beta/N;
    %TWC_beta
    for n=1:n_max_b
        TWC_bet_x(n)=0;
        TWC_bet_y(n)=0;
        Sum_w_beta(n)=0;
        for i=1:N
            TWC_bet_x(n)=TWC_bet_x(n)+w_beta(n,i)*x(i);
            TWC_bet_y(n)=TWC_bet_y(n)+w_beta(n,i)*y(i);
            Sum_w_beta(n)=Sum_w_beta(n)+w_beta(n,i);
        end
        TWC_bet_x(n)=TWC_bet_x(n)/Sum_w_beta(n);
        TWC_bet_y(n)=TWC_bet_y(n)/Sum_w_beta(n);
        CM_x=TWC_bet_x(1);
        CM_y=TWC_bet_y(1);
        d_TWC_CM(n)=sqrt((TWC_bet_x(n)-CM_x)^2+(TWC_bet_y(n)-CM_y)^2);
    figure(1)
    plot(TWC_bet_x(n),TWC_bet_y(n),'.r'),
    drawnow
    end
    for n=2:n_max_b
        if d_TWC_CM(n)>= d_TWC_CM(n-1)& d_TWC_CM(n)>=
d_TWC_CM(n+1)
            n_bet=n
        end
```

```
    end
    d_TWC_CM_bet=max(sqrt((TWC_bet_x-CM_x).^2+(TWC_bet_y-CM_y).^2));
    d_TWC_CM(n_bet);
    bet_str=beta(n_bet);
    %%%%%%%%%%%%%%
    del_gam=input('del_gam= '); %For example: del_gam=0.001
    n_max=input('n_max= '); %For example: n_max=10000
    for n=1:n_max
        gamma(n)=(n-1)*del_gam;
        for i=1:N
            TWC_gam_x(i,n)=0;
            TWC_gam_y(i,n)=0;
            Z(i,n)=0;
            for j=1:N
                P_gam(i,j,n)=exp(-d(i,j)*gamma(n)/Dmax);
                TWC_gam_x(i,n)=TWC_gam_x(i,n)+P_gam(i,j,n)*x(j);
                TWC_gam_y(i,n)=TWC_gam_y(i,n)+P_gam(i,j,n)*y(j);
                Z(i,n)=Z(i,n)+P_gam(i,j,n);
            end
            TWC_gam_x(i,n)=TWC_gam_x(i,n)/Z(i,n);
            TWC_gam_y(i,n)=TWC_gam_y(i,n)/Z(i,n);
            figure(1)
            plot(TWC_gam_x(i,1),TWC_gam_y(i,1),'*r',TWC_gam_x(i,n),
TWC_gam_y(i,n),'.g'),grid on
            drawnow
        end
    end
    %%%%%%%%%%
    del=input('del= '); %For example: del=0.1
    x_p=min(x)-5*del:del:max(x)+20*del;
    y_p=min(y)-5*del:del:max(y)+20*del;
    for i=1:length(x_p)
        for j=1:length(y_p)
            c(i,j)=0;
            for n=1:N
                for t=1:n_max
                    m(i,j,n,t)=sqrt((x_p(i)-TWC_gam_x(n,t))^2+(y_p(j)-
TWC_gam_y(n,t))^2);
                    c(i,j)=c(i,j)+exp(-m(i,j,n,t)*bet_str/Dmax)/n_max;
                end
            end
        end
    end
    for i=1:length(x_p)
        for j=1:length(y_p)
```

```
            xx(i,j)=x_p(i);
            yy(i,j)=y_p(j);
        end
    end
    figure(2)
    contourf(xx,yy,c),title('TWC\gamma')
    figure(3)
    surface(x_p,y_p,c'),title('TWC\gamma')
    figure(4)
    surf(x_p,y_p,c'),title('TWC\gamma')
    t_max=n_max
    i_thet=1;
    TWC_ijt_thet_x=[];
    TWC_ijt_thet_y=[];
    for t=1:t_max
        TWC_ij_thet_x=[];
        TWC_ij_thet_y=[];
        for i=1:N
            for j=1:N
                TWC_thet_x(i,j,t)=x(j)+TWC_gam_x(i,t)-TWC_gam_x(j,t);
                TWC_thet_y(i,j,t)=y(j)+TWC_gam_y(i,t)-TWC_gam_y(j,t);
                TWC_theta_x(i_thet)=TWC_thet_x(i,j,t);
                TWC_theta_y(i_thet)=TWC_thet_y(i,j,t);
                i_thet=i_thet+1;
                TWC_ij_thet_x=[TWC_ij_thet_x,TWC_thet_x(i,j,t)];
                TWC_ij_thet_y=[TWC_ij_thet_y,TWC_thet_y(i,j,t)];
            end
        end
    TWC_ijt_thet_x=[TWC_ijt_thet_x;TWC_ij_thet_x];
    TWC_ijt_thet_y=[TWC_ijt_thet_y;TWC_ij_thet_y];
    end
    figure(5)
    plot(x,y,'*b',TWC_theta_x,TWC_theta_y,'.r'),xlabel('x'),ylabel('y'),
title('TWC\theta')
    G_thet=[];
    for i=1:N
        for j=1:N
            theta(i,j)=0;
            for t=1:t_max-1
                if i~=j
                    theta(i,j)=theta(i,j)+sqrt((TWC_thet_x(i,j,t+1)-
TWC_thet_x(i,j,t))^2+
                    (TWC_thet_y(i,j,t+1)-TWC_thet_y(i,j,t))^2);
                    G_thet=[G_thet;i j theta(i,j)];
                end
```

```
            end
        end
    end
    G_thet_MST=[];
    for i=1:N
        for j=i:N
            if theta(i,j) < theta(j,i)
                theta_MST(i,j)=theta(i,j);
            else
                theta_MST(i,j)=theta(j,i);
                G_thet_MST=[G_thet_MST;i j theta_MST(i,j)];
            end
        end
    end
    G_MST = graph(G_thet_MST(:,1)',G_thet_MST(:,2)',G_thet_MST(:,3)');
    G_MST.Edges
    figure(6)
    p = plot(G_MST,'EdgeLabel',G_MST.Edges.Weight);
    tree = minspantree(G_MST);
    tree.Edges
    highlight(p,tree)
```

Index

M. Buscema et al., *The Topological Weighted Centroid: A New Vision of Geographic Profiling*, Studies in Computational Intelligence 1095,
https://doi.org/10.1007/978-3-031-28901-9